SUPER 8

An Illustrated History
by Danny Plotnick

RARE BIRD

CONTENTS

SUPER 8 IS DEAD! LONG LIVE SUPER 8! 10
WHAT IS SUPER 8? 12
SUPER 8—MY EARLY YEARS 22
CAMERAS 30
FILM STOCKS 54
PROJECTORS 66
EDITING 78
SOUND 88
PROCESSING 96
PRINTS 106
VIDEO TRANSFER 112
EXHIBITION 120
DEATH & REINVENTION 128
INTERVIEWS 134

    THE 1960'S AND 1970s 136
    ROLAND ZAVADA 138
    LENNY LIPTON 141
    JOHN PORTER 145
    JAMES MACKAY ON DEREK JARMAN ... 148
    PAUL SHEPTOW 153
    JONATHAN TYMAN 156
    ROCKY SCHENK 160

    JAMES NARES 164
    BETH B 168
    NARCISA HIRSCH 172
    THE 1980s AND 1990s 174
    RICHARD LINKLATER 176
    PEGGY AHWESH 180
    DAVE MARKEY 183
    G. B. JONES AND BRUCE LABRUCE 187
    MARTHA COLBURN 191
    MATTHIAS MÜLLER 194
    SILT 197
    NORWOOD CHEEK 202
    MELINDA STONE 204
    THE 2000'S AND BEYOND 208
    LISA MARR AND PAOLO DAVANZO 209
    ED SAYERS 213
    KARISSA HAHN 214

THE LEGACY OF SUPER 8 216
NOTES 227
PHOTO CREDITS 231

SUPER 8 IS DEAD! LONG LIVE SUPER 8!

What the hell is this book?

Ever since I picked up a Super 8 camera back in 1985, people have been saying that Super 8 is dead. In a way they're right. By 1985, companies had all but stopped making Super 8 cameras and projectors. But, in a way, all those doomsayers were also wrong. Kodak was still manufacturing Super 8 film, and though no new equipment was rolling off the assembly lines in Rochester, Nagano, or Munich, you could still pick up used equipment or dead stock in a big city or backwater camera shop. Sure, the lab support wasn't as robust as it was for those shooting on 16mm or 35mm film, but you could still carve out a space as an up-and-coming filmmaker using Super 8. You could hone your chops and perfect your craft with this home movie medium—a medium that was being plowed under by the burgeoning home video industry.

But 1985 was a long time ago. All the equipment that wasn't snatched up in 1985 sat in flea markets, pawn shops, and thrift stores, getting older and older, parts decaying and corroding as the years passed. In 2019, you can still pick up used equipment, but that equipment is more than thirty years old. In the intervening decades, Kodak continued to eliminate film stocks, and in many respects, Super 8 is even more dead in 2019 than it was in 1985. But true believers continue to shoot Super 8 and craft beautiful images out of this beleaguered but beloved film gauge.

This book looks back at the history of Super 8, explores how the medium has evolved through the years, and provides a glimpse into what it was like to be a no-budget artist in the pre-digital age. Filmmaking was expensive business and Super 8 offered a low-cost alternative for young people wanting to use film to say something about the times in which they lived. How did they go about it when there was still plenty of equipment to be found, and enough labs to support their filmmaking practice? In particular, this book explores the challenge of making sync sound films in Super 8, a methodology that took a major hit in 1996 when Kodak discontinued sound Super 8 film.

In 2015, Super 8 celebrated its fiftieth birthday, bringing it that much closer to death. But don't worry, as much as Super 8 is pretty much dead, it's still very much alive. Kodak plans to unveil a new Super 8 camera in 2019, possibly breathing new life into a decaying body. Dead. But not quite dead.

Danny Plotnick, 2019

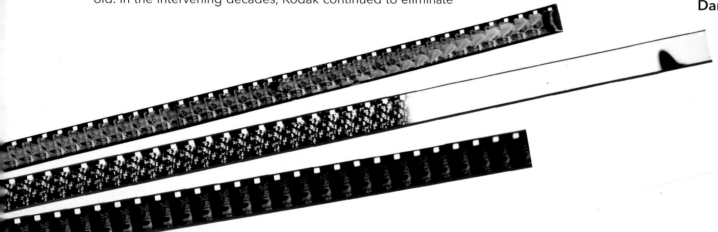

CHAPTER 1:

WHAT IS SUPER 8?

WHAT IS SUPER 8?

You can talk about Super 8 from a purely technological perspective.

The Super 8 motion picture format was developed by Kodak in 1965. Super 8 has smaller sprockets than its precursor, Regular 8 (sometimes referred to as Standard 8), which means the Super 8 image is spread across a larger swath of its 8mm celluloid strip than Regular 8. As a result, the Super 8 image is almost 50 percent larger than the 8mm image, therefore containing more information and producing an image with greater definition than its predecessor. At the time, the other movie format on the block that was accessible to the general public was 16mm, which, by virtue of its 16mm-wide celluloid palette, contained more than three times the information of either of the 8mm formats.

You can also talk about how Super 8 fits into the continuum of film history. Both 16mm and Regular 8mm had been around for decades prior to Super 8's release. The 16mm format was born in 1923; Regular 8 followed in 1932. Throughout the years, 16mm was used for a variety of purposes. Until the release of Regular 8 in the 1930s, 16mm was the principal home movie format of choice in the US. In the 1940s, it was used by combat cameramen in WWII. In the 1950s, it found widespread use in the educational, television news, and industrial film markets. In the 1960s, it became the favored gauge of documentary, experimental, and independent filmmakers from Cassavetes to Godard to Pennebaker to Brakhage. Shooting in 16mm offered a beautiful image at a fraction of the cost of 35mm film.

Meanwhile, Regular 8 solidified itself as the stalwart of the home movie world. While the image quality of Regular 8 didn't pop as much as 16mm, it was a substantially cheaper format. The 8mm cameras, projectors, editing gear, and film stock were a fraction of the cost of 16mm equipment and accessories. This affordability made it the format of choice for dads wanting to document family birthdays, vacations, and graduations.

Upon its arrival, Super 8 quickly dominated the home movie market, supplanting Regular 8 for market share. Not only did it have a superior image compared to Regular 8, but it was much simpler to operate. Regular 8 was a tricky beast. It came in 25-foot rolls that were 16mm wide that you needed to manually thread into your camera. Because the camera's film gate was only 8mm wide, when you shot, you only exposed 8mm, or half of the film strip. You shot for twenty-five feet (roughly 1 minute and 50 seconds), at which point you needed to pull the film out of the camera, flip it over, rethread it, and then shoot the 25-foot roll a second time, this time exposing the other 8mm of film. You would ship the finished reel to a lab for processing where they split the film lengthwise and spliced the two twenty-five-foot sections together, giving you a fifty-foot film reel, approximately 3 minutes and 40 seconds in length, and 8mm wide.

Super 8 was significantly less cumbersome. The film came in a 50-foot, light-tight cartridge that simply popped into the camera's film chamber. It took a mere matter of seconds to insert the cartridge, and, presto, you were ready to shoot. That process was no more difficult than popping videotape into a camcorder. No muss, no fuss. Additionally, there was no significant price difference between Super 8 and Regular 8. Super 8 had a better image and was easier to use. It was an easy choice to make for home-movie enthusiasts. Kodak and Bell & Howell were the first manufacturers to jump into the Super 8 pool, and even they "were surprised at how well the new format caught on…and how quickly it displaced Standard 8mm in the home market." As a result of the new gauge's success, Kodak discontinued all but one Standard 8mm model by the end of 1966.

L-R: Regular 8, Super 8, and 16mm.

Interestingly, though Super 8 took the home-movie world by storm, the new format was born out of Kodak's desire to find a better format for "educational, industrial, and commercial purposes." Kodak believed a better quality format would make inroads into these markets, which didn't traditionally work in 8mm, instead producing and distributing mostly on 16mm. Generating 16mm prints was expensive, so companies would sometimes create 8mm reduction prints to save money. The 8mm prints, however, were not always embraced due to the limited information carried by such a small image. Super 8's increased image area was designed with an eye on improving the overall quality of reduction prints for the boardroom and the classroom. Super 8 also promised additional savings by lowering the cost of reduction prints. Its dimensions made it "possible to make four Super 8 prints at one time on 35mm wide film." This would be a boon for education and business markets, at least that's how the thinking went.

In a particularly savvy business move, Kodak made the industry aware of its intentions. The company "recognized that it would be of significant benefit to the industry if the new format were internationally adopted and standardized," so Kodak handed over its plans to equipment manufacturers worldwide. This move paid huge dividends and helped fuel the format's rapid growth.

As early as 1966, thirteen companies had begun producing Super 8 cameras. That number ballooned to nearly thirty companies by 1967. Ultimately, more than eighty companies would go on to manufacture Super 8 cameras. In addition to Kodak and Bell & Howell, Canon, Nikon, Yashica, Sankyo, Argus, GAF, Bauer, and Braun were just some of the companies entering the fray. No two cameras were alike. Some were high-end, some were low-end. Some cameras were built like tanks, while others were riddled with cheapo, plastic parts. Some had fantastic lenses with a wide range of zoom capability, while others had rinky-dink, fixed-focus lenses. Some cameras featured a large array of manual controls, while others contained only automatic functions. All of the cameras shot film at 18 frames per second (fps), which was not the 24 fps industry standard. Only some of the higher end cameras filmed at 24 fps.

With so many manufacturers producing cameras, there was no consistent set of functions on the equipment that began to flood the market. This may have contributed to the feeling that there was no clear-cut production methodology built up around the format. This lack of standardization created a space in which amateurists could develop their own style, and as a result, Super 8 took on different meanings for different people.

At the outset, Super 8 production was the purview of dads documenting their family's Kodak moments. Due to its proliferation amongst suburban families, Super 8 quickly fell into the hands of teenagers, who took to making backyard monster movies, parroting the no-budget style of Roger Corman and his ilk. The likes of Sam Raimi, J. J. Abrams, and Vince Gilligan cut their teeth this way.

By the early 1970s, realizing that they could work at a fraction of the cost of 16mm, independent, experimental, and underground filmmakers soon got in on the action. Super

Super 8

Regular 8

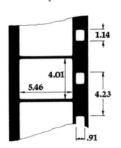

Super 8

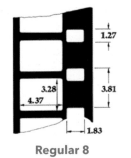

Regular 8

NEW SUPER 8 KODACHROME II FILM OFFERS SCREEN IMAGES
THAT ARE BRIGHTER, SHARPER, MORE STABLE

NEW YORK, April 26 -- A new Super 8 movie film, pioneered

developed by Eastman Kodak to provide 50 percent more image area

regular 8mm film, was shown here today publicly for the first

as the company introduced Kodak Instamatic Movie Cameras and

ctors for use with the film.

Kodak Super 8 film is an improved Kodachrome II, Type A,

will be supplied in Kodapak Movie Cartridges for quick, non-thread

ding into the new cameras. The film is for indoor and outdoor use

h provision for automatic control of filter and ASA film speed

tting built into the cartridge and cameras. The cartridges contain

0 feet of the 8mm wide film and do not have to be turned over; the

camera user can shoot 50 feet without interruption.

While today's showing in the Waldorf-Astoria Hotel was

the first public introduction of the film, Kodak from time to time

d other photographic manufacturers in this country and

8 was becoming an international movement. Lenny Lipton dabbled in his singular brand of documentary in the Bay Area; Derek Jarman made stunning films set amidst the avant-garde world of London; Narcisa Hirsch & Claudio Caldini created experimental work in Argentina, and by the early 1980s, young filmmakers involved in the German film organization Gegenlicht embraced Super 8 as their medium of choice to document the anti-nuclear movement.

From 1965 to 1972, films had to be shot silently. Accompanying soundtracks, if any, would in all likelihood be recorded onto reel-to-reel or cassette tapes to be played simultaneously with the film's projection. In 1972, sound was born. Renowned documentarian, Richard Leacock worked with MIT to develop a double system sound, Super 8 workflow. This process allowed Super 8 to function like 16mm. A Nizo camera ran alongside a Sony cassette recorder, both synced at a precise 24 frames per second. The sound and picture were married in post. Editing was done on a Super 8 flatbed; the entire process was akin to 16mm. Down the street at Harvard, Bob Doyle was also working on a double system setup of his own. His system, which followed in 1973, led him to start Super8 Sound, a company that was a critical player in the world of Super 8.

Double system allowed Super 8 to function in a more traditional and professional manner. The hope was that this workflow would be appealing to both documentarians and television news-gathering services. It had the familiarity of 16mm production, but at a reduced price point. While there was a modicum of work produced in this way, double system Super 8 never became a dominant force. The cost of the initial Leacock/MIT system pushed $7,000, and didn't hold sync particularly well. Doyle's system, which was much cheaper and held its sync, won out. Super8 Sound did an incredibly brisk business, selling 2,000 units by 1976. This approach was certainly cheaper than 16mm, but with video technology just over the horizon, it was a proposition with little future. By 1980, very little Super 8 work was being produced in this fashion.

The sound format that revolutionized Super 8 production was developed by Kodak in 1973. The company introduced a way to shoot single system sound with their new sound Super 8 film stocks and Ektasound 130 and 140 cameras. All you had to do was plug a microphone straight into your camera, start shooting, and you'd be recording sync sound directly onto your film. This was truly a game-changer. Due to its simplicity, single system sound Super 8 opened up myriad possibilities for filmmakers. The sound film had a .027mm magnetic audio stripe, similar to the iron oxide tape used in audio cassettes, adhered to the edge of the film opposite the sprockets. Many filmmakers used this feature to capture sync sound on the set. If you shot silently, you could now add sound in postproduction by lining a mic, turntable, or cassette deck into a sound projector that could record onto the magnetic stripe. The film also had a second smaller stripe on its other edge, which could be accessed to add additional sounds in postproduction. Kodak's Ektasound 130 camera came in at a reasonable $189.50, making it a bargain compared to the more expensive double system universe.

Bolstered by the advent of sound film, Super 8 was embraced by a new breed of filmmaker. Combining their love of music and

Canon 312 XL-S, 1977, $372

the French New Wave, the No Wave Cinema movement was born in New York City in the mid-1970s. Jim Jarmusch, Amos Poe, James Nares, John Lurie, Charlie Ahearn, Vivienne Dick, and Beth and Scott B were just a few of the folks documenting the gritty, urban lifestyle of their bankrupt metropolis. By the 1980s, this scene gave way to the harder edged Cinema of Transgression featuring the likes of Richard Kern and Nick Zedd.

By the mid-1970s, Super 8 also began infiltrating the college ranks with over eighty American universities offering Super 8 filmmaking courses. Due to its low cost, many universities embraced Super 8, teaching introductory production classes with the format. Super 8 provided an easier introduction to the filmmaking process than 16mm. Super 8's easy-to-load cartridges, built-in automatic light meters, and ability to record high quality sync sound were just some of the features that made it a good format for beginners. Additionally, if you had access to mid-range or high-end cameras with nice lenses, the image quality was quite sharp. The low cost of equipment, film, and processing appealed to both university production programs and the broke college students who filled them.

Media arts centers, which began dotting the landscape in the 1970s, also loved the low barrier to entry that Super 8 provided. Many young people who chose to forego the college experience received film training at these art hubs, or simply learned the medium on their own. Richard Linklater (*Dazed and Confused*, *Boyhood*) chose not to go to college for film. Instead, he taught himself the filmmaking process and made his first feature on Super 8 before he made *Slacker*. While it might be a stretch to claim Super 8 had an influence on the rising independent film movement that took off in the late 1980s, there's no doubt many independent filmmakers, such as Todd Haynes, Spike Lee, Wes Anderson, and Alex Gibney, framed their first shots while looking through the viewfinder of a Super 8 camera.

Super 8 held great promise, but with the rise of video in the late 1970s and early 1980s, the bottom quickly fell out of the Super 8 market. "Bell & Howell closed its Consumer Division in 1979" and Kodak stopped manufacturing Super 8 cameras in 1981. By 1985 almost no new equipment was being manufactured. As the years progressed, fewer services became available to Super 8 filmmakers and beloved film stocks were discontinued. However, these obstacles didn't stop young people from picking up thrifted gear and trying their hand in Super 8.

Why did they pick up this discarded equipment? Because there was a story they wanted to tell. Even though video was on the rise, it didn't look like film. If you wanted to produce something filmic, you had to shoot film. And Super 8, though it was often seen as a second-class citizen, was still film. If you had a good eye, a good camera, and embraced the format's limitations, you could make strong work.

Due to its economical and egalitarian nature, no two Super 8 filmmakers approached their craft in the same way. Its colorful and somewhat ragtag story is told best through the personal experiences of those who dared to create something grand with a home-movie medium. Could it be done? Of course it could, but it was not easy. Super 8 could be beautiful, but it was a finicky beast. As you turn the pages of this book, you will hear many of the adventures from Super 8 pioneers who aimed for perfection in an imperfect medium. ■

Prototype of Kodak M4, 1965–1967, $69.50

CHAPTER 2:

SUPER 8: MY EARLY YEARS

SUPER 8: MY EARLY YEARS

I started making Super 8 films by chance.

It was the mid-1980s. I was in college, and I was on my way to being a history major. In a Russian history class, we viewed Sergei Eisenstein's *Strike*. At the film's climax, Eisenstein, ever the master editor, cross cut workers being gunned down by government thugs with images of cows being slaughtered. This was revelatory. I had never seen anything like that before in a film. It was rebellious. It dripped with anti-authoritarian sentiment. It was a style of filmmaking far removed from anything I had previously witnessed. It had a punk rock vibe. At least that was my takeaway. At the time, I was involved in the underground music scene, and this audacious moment in *Strike* struck a chord with me. I went back and watched the film two more times that week. Over the course of those viewings, I stopped wanting to be a history major and decided to become a filmmaker. I wanted to make art like that. Art that made you sit up in your seat and take notice. Eisenstein changed my life.

I was an undergraduate at the University of Michigan, one of the top liberal arts schools in the country. Unfortunately for me, the film department was, arguably, the worst department in the school. There were only three official classes: a film history course; a video production class where you lugged around backbreaking 3/4" video Portapaks; and a Super 8 production class. Not only was the class selection limited, but the amount of equipment available was scant, and the quality of that gear was gamey, at best.

I have no recollection of the kinds of cameras we used, though I can guarantee you we weren't using the top-of-the-line Brauns and Beaulieus. There was one projector available to the entire program and it was silent. We could, and did, make sound films—however, we were only able to watch them when the professor brought his own sound projector to class.

This lack of postproduction sound equipment made cutting sound films nearly impossible. You had to befriend the teacher and convince him to let you borrow his equipment. He was actually nice enough about it, but I do remember some awkward trips to meet him at various locales around town to gather up the gear. When you're nineteen, it feels strange to interrupt your professor's adult-league softball game to borrow film equipment. You follow him to his car, parked in the residential section of a college town, the trunk pops open, an exchange is made. As odd as it was, it also felt strangely adult. Like you were taking control of your own life, stepping out of your comfort zone to get what you needed to make your art.

Eventually, one of my classmates bought his own projector. I used it a lot. And, perhaps, as a result, I starred in a horror movie of his. Was I interested in horror movies? No. Did I consider myself an actor? No. But that was the price of gaining access to a sound projector, which I needed to edit and watch rough cuts of my own movies.

Was I the hero of that horror movie? A hapless, butchered victim? I'm unsure, though I'm pretty certain I don't want to head down into the archive and unearth that film. I do remember a scene of me leaning against the wall in a dorm room and looking up at a poster of Farrah Fawcett. Or maybe it was a more mid-1980s appropriate pin-up. Either way, I shudder just thinking about it. The point is, if you didn't have adequate resources of your own, this was simply the price you had to pay if you wanted to make movies.

I soon bought a projector of my very own—a Chinon SP 330 MV. I bought it at a suburban Ritz Camera. They were a pretty stinky photography chain at the time and though I never really liked the place, it's where I bought my first, and only, projector. It cost just over $100. It was a big purchase for me, but a necessary one. No longer would I be dependent on my professor or my horror-obsessed friend. The phrase DIY was not in use yet, but unwittingly I, along with many other Super 8 filmmakers, was heading down that path.

I could borrow cameras from school. I could borrow a better camera, a GAF 805 M, from my friend Mike. I could get the editing supplies from school. With my own projector in hand, I could now watch, or better yet exhibit, the finished product.

It was one thing to make a film, but if you wanted to have an impact, the way Eisenstein impacted me, you had to show your films to an audience. This was no easy task in the mid-1980s. There was no YouTube. No internet. No home for short films on TV. Most college kids didn't even have VCRs yet. Festivals and public-access TV were the only outlets likely to show your films. But accessing those venues was a challenge. As a nascent filmmaker, it was hard enough to make a film. I barely knew what I was doing behind the camera, so there was no way that I had any idea how to access the festival circuit. Plus, my films were too raw, too unfinished to enter that realm. But I was passionate about the work I was doing. Perhaps I was misguided, too swayed by my youthful enthusiasm, too enamored by my own reflection, but I thought that like-minded folks would respond to my films. Sure, they were rough around the edges, but so was punk rock!

The author in his youth.

With a projector now part of my arsenal, I realized that when I was done with a film I could invite friends to my house, and have a screening in the living room. The projector was lightweight, which meant I could be mobile. I could take it to a friend's house and show movies in the backyard after a barbeque. I had friends in bands. I could ask them if I could show a film before their sets. I could pack up that projector, set it up on a table in a bar and let the show begin. I wasn't going to wait for the people to find me. I was gonna take the show to the people. I was taking matters of distribution into my own hands. I was making it up as I went. Honestly, I had no idea what I was doing. I was learning on the fly.

Why did I choose Super 8? At the time, the prevailing reason to use Super 8 was because it was significantly cheaper than 16mm or 35mm. But that wasn't my motivation. I chose Super 8 because that was the format of choice in the class that my college offered. I didn't know any other world or what any other options were. No one ever explained to me the difference between any of the formats. Had U of M offered a 16mm class, I'm sure I would have forged a career with those tools. Had there only been a 3/4" video class, I'm sure that would have been my gauge of choice. Simply put, I didn't know any better. I just took what was in front of me and ran with it. ∎

Chinon 330 MV, 1980, $199.50

CASE STUDY

Skate Witches

OF ALL MY FILMS, *Skate Witches* is the one that has thrived in the Internet Age. This is somewhat fascinating and perplexing to me. The film was made in the space of an afternoon in Ann Arbor, Michigan, in 1986. The plot involves a gang of female skateboarders who terrorize the boy skateboarders in town. Though their skating leaves a bit to be desired, they talk a mean game, push boy skateboarders off their boards, have pet rats, and sport bad-ass leather jackets with their crew name stenciled in white paint. Did I mention the film clocks in at just under two minutes? I imagine that has something to do with its internet success.

What fascinates me is that this was one of my first films. As will become evident as you leaf through this book, Super 8 was an unforgiving medium. For someone like me, who was self-taught, unforgiving meant that the early films were choc-a-block full of mistakes. *Skate Witches* is riddled with problems. When I taught Super 8 classes, this was one of the first films I showed. I did this not as a back-patting power move, but to show off both how beautiful Super 8 could be if shot well, and to demonstrate how easy it was to fall victim to Super 8's shortcomings.

Perhaps I should start with the beauty. *Skate Witches* was shot on Kodachrome 40 on a bright, sunny summer day. This combination meant that the blue sky popped. The rich hues of the green trees were on display in their full, Midwestern bloom. The red-shingled rooftops shimmered, and the flesh tones were vibrant and pink-flushed. *Skate Witches* really did show off Kodachrome's hyper-saturated color palette. The video transfers the film received over the years have never done the color true justice. To see the beauty of the Kodachrome, you really need to be looking at a film projection of the Kodachrome original.

But the film has its problems as well. Problems a first-time filmmaker would encounter. The film opens with a blast of hardcore. Faction's "Skate and Destroy" shreds over the film's opening credit, a shot of the Skate Witches logo emblazoned across one of the witch's leather jackets. The song was pulled from a cassette comp the Queen Witch had. To get it onto the film, I lined a cassette player into my projector. The projector had the ability to record sound directly onto the film's magnetic audio stripe. The only problem being that the projector's record button made a god-awful electronic squawking whenever you engaged or disengaged it. That wasn't a problem on the front end of the film because I pressed record while the film's silent leader was passing through the projector. The problem was when I pressed stop on the record button in the middle

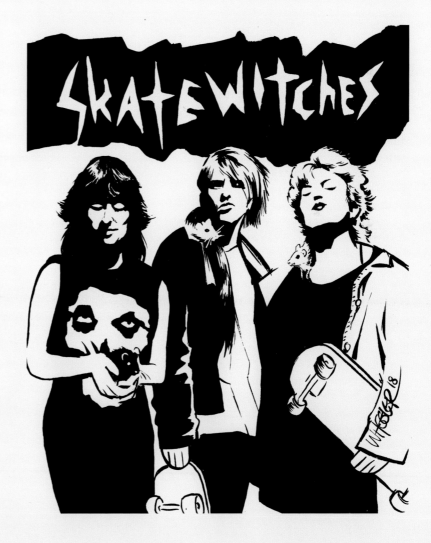

27

CASE STUDY

of the first scene of the film. An unholy screeching, like someone pushing a needle across a record, was forever tattooed on the film's sound track. This taught me a lot about what not to do in the future, but in that moment, the unforgiving nature of Super 8 reared its ugly head and as a result that sound forever remains part of the film's soundtrack. In the digital universe, I've had several opportunities to clean up that mess, but I've always chosen to leave it. It's an authentic mistake that has been part of the film for over thirty years. Why fix it now? Strangely, on the internet, where many great things and many horrific things have been said about the film, no one has ever complained about that fucked-up bit of audio. That sound fart drives me nuts, but the internet hoards couldn't care less. Go figure! I imagine if I cleaned it up, someone would inevitably complain about that!

The other mistake that you can find in every shot is a pulsing light meter. All Super 8 cameras have a built-in automatic light meter. I hadn't quite gotten the hang of the best way to use that back in 1986. You would engage the light meter by pressing the camera's trigger in halfway. When you got to the halfway point, it would take the meter a second to stabilize and find the right f-stop reading. Once the meter stabilized you were ready to shoot. Sadly, that's not how I was rocking things back then. To conserve film, I would yell "Action!" and the second the Skate Witches started delivering lines, I'd press the trigger in full-bore. This meant that the light meter was hunting for the f-stop as the action was unfolding. At the head of each shot of the film, you can see this pulsing take place. Live and learn. And I did. I never made that mistake again, but it lives on in *Skate Witches*. Again, nobody ever complains about that mess either.

The other strange development is that though I spent a lot of money getting quality Super 8 transfers of my films, the version of *Skate Witches* that has caused an internet ruckus is a compromised transfer. In the 1990s I transferred the film to 1", a high-end video stock. When 1" lost favor, the transfer was duped to Beta SP, another high-end tape stock. When it came time to move into the digital world in the early 2000s, I dumped all my Beta SP down to MiniDV and brought the films into the computer world from the MiniDV tapes. Even with these generational losses, the high-quality transfers still looked good. But for some reason, *Skate Witches* never made it to the MiniDV from Beta. I don't know why. So when I digitized my films I had two choices. I could pay a transfer house to digitize the Beta or I could digitize the film myself using a VHS dub I had of the original 1" transfer. Digitizing the Beta would have cost me between $60 and $100. Being the cheapskate that I am, I opted to go the VHS route. In my defense, I had no idea of the potential of this thing called YouTube. I had no idea that the film would ultimately garner almost 500,000 views. Had I known that, I would maybe have paid the $60, but at the time it seemed like a waste of money.

The generational loss from 1" to the lo-fi VHS was significant. The VHS had a fuzzy quality to start with. Combine that with YouTube's putrid compression settings in the mid-2000s and the result is film quality that I cringe at whenever I see it on YouTube. But again, nobody seems to complain. They all think it adds to its 1980s charm. They think it's an intentional artifact. They have no idea how I die a bit inside each time I see it.

> **"In my defense, I had no idea of the potential of this thing called YouTube. I had no idea that the film would ultimately garner almost 500,000 views."**

Another reason I chose to not spend the money on digitizing *Skate Witches* is that during the period when I was actively producing Super 8 films, people liked *Skate Witches* well enough, but it was never the film people went gaga over. When it was released, there wasn't even a market for this type of film. In 1986, festivals and cinematheques were the venues that showed short and experimental films. Those venues gravitated toward a more conventional interpretation of experimental filmmaking. *Skate Witches* was not that. It had a foot firmly planted in the mosh pit. While there was precedent for this type of film from the No Wave years, unless you were from New York or an established "art star," it was harder for films of this nature to gain any traction.

Ann Arbor, where the film was shot, had one of the top 8mm film festivals in the country and, I'm not making this up, *Skate Witches* was rejected from the festival. To add insult to injury, each year the festival gave away a prize to the best film made in Michigan. That year, they gave the award to a filmmaker from Ohio! In their estimation, they didn't find one worthy Michigan-made Super 8 film. Giving that award to someone from Ohio was tantamount to fighting words for a Michigander. It would be a couple of years, but that festival would eventually come around to my work. I would ultimately win a "Best in Fest" in 1990 for my film *Steel Belted Romeos*, but that was four years after *Skate Witches* was

CASE STUDY

The hyper-saturated colors of Kodachrome 40 on display in Skate Witches.

rejected—the point being that films with attitude had trouble finding a home in the mid-1980s.

Underground films with this kind of aesthetic would ultimately find a following, but that didn't really take off until the explosion of the underground film festival scene, which kicked off with the birth of the New York Underground Film Festival in 1994. So films like *Skate Witches*, made in the nether regions between the end of No Wave and the beginning of the new American underground, wandered aimlessly through the film landscape.

In short, I never saw the reverence for *Skate Witches* coming. There were hints of excitement around the film in the late 1990s with the birth of skate film festivals like the Mid-Atlantic Skate & Sound Film Festival, and art shows like the San Francisco Arts Commission's "Switch Stance" show, which gave skate and surf culture a serious nod. However, these were fleeting romances.

My ability as a prognosticator has never been great. I certainly couldn't foresee the power of the internet and its capacity to transform how we view films and access the past. I didn't anticipate legions of people using the internet as a thrift store, to unearth archeological treasures from bygone eras. As a result, I couldn't be bothered to make a $60 investment to archive my film in a pristine fashion. Maybe I had my head buried too deep in the analog universe to foresee the future. But people were still entrenched in an analog world, so forgive my blind spot. Super 8 filmmakers spent careers being ignored and passed over, so there was no reason to think this new medium would change that equation. And so, I put up a crappy looking version of a film online. Just my luck that film would be the one to take off. The irony is rich. Though, in a bizarre way, maybe adding a lo-fi VHS sheen to a lo-fi Super 8 film was the secret ingredient to internet fame and fortune! ■

CHAPTER 3:

CAMERAS

CAMERAS

Super 8 cameras are cool—small and sleek with a snazzy pistol grip.

They feel good in your hand. They're lightweight and they cozy up to your eye just right—ideal for run-and-gun guerrilla filmmaking. You can gaffe them to a skateboard or take them into the pit at a hardcore show. They are unobtrusive. They aren't intimidating. You can get intimate with your subject, unlike 16mm which is bulky. You can see it coming a mile away. Not to mention, 16mm is complex with its external light meters, byzantine threading patterns, and changing bags. Super 8 is smart and simple. Pop that 50-foot cartridge into the camera and you're ready to roll in a matter of seconds. Do a quick zoom, grab your focus, take a quick meter reading with the built-in light meter, lock your exposure, and roll. Fifty feet of film fly by in the blink of an eye.

Not too terribly long ago, there were some rock-solid Super 8 cameras. The Canon 1014 XL-S was the primo sound camera. Decked out with a mighty zoom lens (6.5–65mm, f1.4), an intervalometer for time lapse, a variable shutter, and sound recording functions, this camera had it all. Recording sound was as simple as plugging a mic into the mini input on the camera. Voilà, you had sync sound filmmaking at your fingertips, something conventional 16mm production couldn't offer.

The Nikon R 10 was the top of the pops when it came to silent cameras, boasting an equally impressive lens (7–70mm, f1.4) and the ability to do longer in-camera dissolves (100 frames) than the Canon 1014 XL-S (90 frames). In-camera dissolves! That was madness. In a postproduction universe that was reliant on straight cuts, in-camera dissolves added major production value to your film. Not only were both of these cameras kicking out beautiful images, but they both had a heft. You could probably crack someone's skull open with one of those and your camera would still remain functional.

Though they were Japanese companies, the Canons and Nikons were readily available in America. Over in Europe, Beaulieu and Braun designed beautiful cameras, featuring streamlined European craftsmanship and great glass. The Beaulieu 2008 S, 4008 ZM series, 5008, 6008, 7008, and 9008 even allowed for interchangeable C-mount lenses, accommodating most 16mm lenses on the market. The 4008ZM II came "equipped with a 6 to 66mm Schneider Optivaron f/1.8 zoom lens." Schneider was one of the top lens manufacturers out there. Even today, my students stare in awe of the beauty of this camera. As Lenny Lipton astutely observed back in 1975, "It's a fine French machine, looking more like Barbarella's ray gun than a Super 8 camera."

No question about it, the images these dandies produced were outstanding. All of these companies had American distributors, and based on the interviews in this book, the Nizo line of cameras produced by Braun had an incredible reach and were beloved by filmmakers worldwide. The form and function of the Nizos were unparalleled. Founded in the 1920s, Nizo was a German manufacturer of movie cameras. The company "nearly went bankrupt" in the late 1950s and was "taken over by Braun in 1962." Non-filmmakers will certainly know Braun for their razors and kitchen appliances. Braun was a company that took its industrial design seriously and the team at Braun was headed up by renowned designer Dieter Rams. There were many beautiful Super 8 cameras produced, but Rams and Marcel Beaulieu were the only designers whose names you heard talked about in revered tones in the Super 8 world.

There were less rarified, but eminently solid, workhorse cameras as well. By 1967, there were already over seventy models of Super 8 cameras one could purchase. As the 1970s dawned,

Kodak Ekatasound 240, 1976–1979, $399.50

manufacturers like Agfa, Chinon, Minolta, Elmo, and Eumig helped flood the market. Over the lifespan of Super 8, there were over 1,600 models available to consumers worldwide.

Canon had a huge array of cameras, and many of their mid-range cameras were not only excellent, but easy to find. Often found at bargain-basement prices compared to the top-shelf competition listed above, many of these models housed very nice lenses, though with a narrower zoom range than the likes of the 1014 XL-S. The Canon 514 XL-S was an incredibly popular mid-range camera, sporting a 9-45mm zoom. Many of these mid-range cameras also offered sound functionality. They may not have had all the bells and whistles of their top-end brethren, but they got the job done.

By 1985, with the exception of Beaulieu, companies stopped producing Super 8 cameras. You could still find them throughout the eighties and nineties by directing your attention toward fancy camera stores, scouring the classifieds, or visiting camera swaps, flea markets, and pawn shops. After all, more than 10,000,000 Super 8 cameras had been sold to households in the US before the Super 8 market crashed, so there were still plenty of used cameras to be found.

All Super 8 cameras ran on AA batteries. Any shopping excursion necessitated bringing a pack of AAs. If you found a camera, saw that there were no obvious dents on the body or on the lens, and if the AAs powered it up, it was probably worth the risk.

There was no standard function set on a Super 8 camera. Many of the low-end cameras had almost no functions other than a zoom lens. So, what was the range of options you might find on a Super 8?

Sound
I suppose the first question you had to ask yourself if you were on the hunt for a Super 8 camera was, "Do I want to make sound films?" If the answer was yes, then you needed a sound camera. There were both silent and sound models of Super 8 cameras, and assessing whether or not you were looking at a sound camera was easy to figure out. The sound models contained a sound head in the film chamber. There was no mistaking that component. The exterior of the cameras featured a mini-input jack for the mic and a headphone jack for sound monitoring. Most of the sound cameras utilized auto levels, though a handful (like the 1014 XL-S) offered some ability to control levels.

Zoom
A nice zoom was essential. Almost all Super 8 cameras came with fixed-mount lenses. No interchanging of lenses went on other than a handful of exceptions like those noted above. What was on the camera was what you got, so a nice zoom gave you options. The greater zoom range you had, the better off you would be. The range was all over the map, with a 12x zoom being the max. If you got to a 10x zoom, you were doing well for yourself. Most cameras had an auto zoom for nice, smooth zooming. For auto zooming, many cameras had one default speed, while some of the fancier cameras had variable speed zoom settings.

Manual Exposure
Manual exposure was a necessity in my book, though many cameras only came outfitted with automatic exposure. For Super 8 metering, you would use the auto exposure to serve as the basis for your final exposure. You could override the auto exposure and dial in an f-stop if you were in lighting conditions that proved confusing to the camera's reflected-light meter. If you had manual control, you could also lock your exposure in place to ensure that your exposure wouldn't fluctuate over the course of your shot.

Many cheapo cameras only had auto exposure with no manual overrides. Some cameras had primarily auto exposure functionality, but with a couple of workarounds. A BLC (back light control) button was commonly found on cameras, allowing the iris to open an additional stop or two if you deemed it necessary. It was used to address backlit situations, but obviously you could engage it whenever you wanted.

Some cameras also came with an EE (electronic eye) or AE (auto exposure) lock. By holding this button down while shooting, you could lock your automatically generated

Canon 1014 XL-S, 1979–1983, $1,134

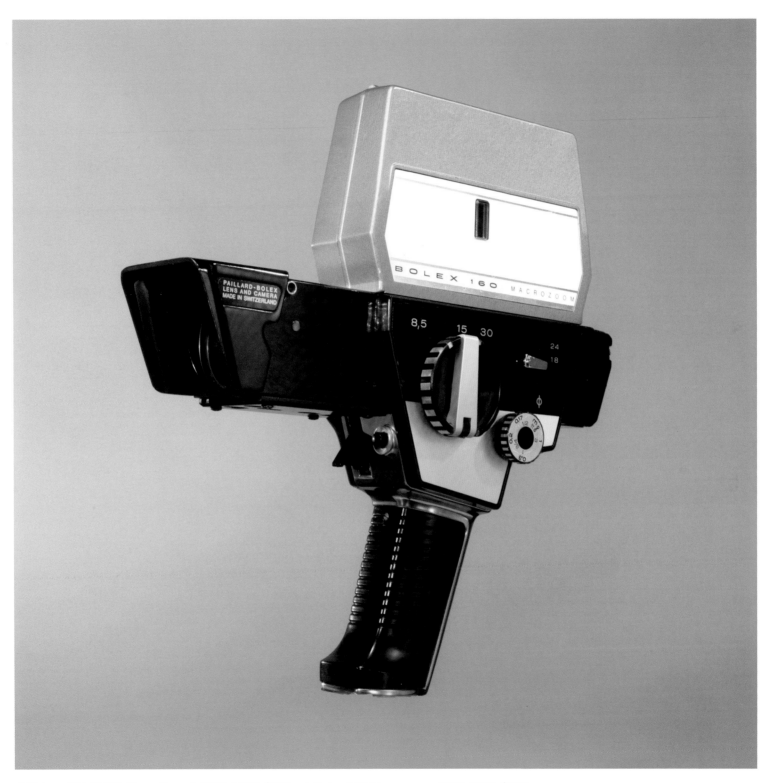

◀ Chinon 506 SM XL Direct Sound, 1976–1978, $520   ▲ Bolex 160 Macrozoom, 1970–1971, $300

Kodak XL 320, 1974–1981
Our Gang XL 320 Packaging, 1978, $99.50

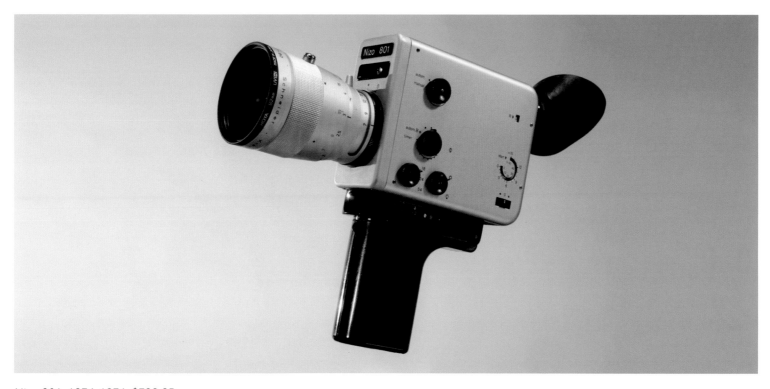

Nizo 801, 1974–1976, $728.95

Canon Auto Zoom 1218, 1968–1974, $800

Kodak Ektasound 160, 1974–1977, $398.50

exposure into place. This was ideal if your light values would be changing over the course of the shot. However, this was quite awkward for a shooter, because you needed one finger to hold down the trigger and another finger to hold the exposure lock in place. But this was a lifesaver for cameras without true manual control of exposure. The Canon 514 XL-S was one of the most popular of all Super 8 cameras, and this was the exposure situation on that camera.

Speed

All Super 8s shot at 18 fps, which was not the industry standard, given that 16mm and 35mm shot at 24 fps. Ultimately, shooting at 18 fps was more cost-effective. A 50-foot cartridge of film lasted 3:20 at 18 fps and 2:30 at 24 fps, so you got more film for your money at 18 fps. If you were planning to blow your film up to 16mm, shooting at 18 fps was not recommended since the frame rates wouldn't match. If you were shooting subjects with a lot of motion, 18 fps was not recommended. Beyond that, 18 fps got the job done at a fraction of the cost.

Many Super 8 cameras did allow you to shoot 24 fps, as well as other frame rates. Some cameras had fast motion and slow motion pre-sets. Many cameras had single frame functionality for animation, and a couple like the Canon 1014 XL-S and Nizo S 560 had intervalometers, which allowed you to shoot time lapse.

Filters

The majority of Super 8 stocks were tungsten balanced and the cameras came with two basic filter settings to account for whether you were shooting with daylight or with tungsten light. On some cameras the filter switch was easy to comprehend. A sun symbol indicated proper color balance when shooting in daylight. A light bulb symbol indicated proper color balance when shooting with tungsten lights. On some cameras there was no easy-to-decipher switch. Some cameras had a movie light plug where a screw needed to be inserted into a thread on top of the camera when shooting with tungsten light—the principal being that in the home movie world, you could screw a light into that mount in order to illuminate your scene. By screwing the light in, you would trigger a mechanical train that would retract the daylight filter (type 85A) housed in the body of the camera, allowing you to shoot in tungsten light without fear. Other cameras had a filter key that operated in the same manner as the light plug.

Some cameras also came with ND (neutral density) filters. Not many did, but what a nice option this was to help prevent overexposure on sunny days.

Fades/Dissolve

Now we're moving into rarified territory. A handful of cameras did auto fades. An even smaller subset did dissolves at the push of a button.

Variable Shutters

A scant few had a variable shutter, which allowed you to open the camera's shutter to a wider angle, allowing more light to get to the film gate. This was great if you knew you would be operating in low-light situations. It could buy you a stop, saving your film in certain situations.

200-foot Cartridge

In 1975, Kodak unleased the Supermatic 200 camera, which accepted a 200-foot cartridge. Kodak produced a handful of sound and silent stocks in the 200-foot configuration. All told, about thirty models accepted these larger cartridges including a smattering of Elmo, Nizo, Chinon, Cosina, Beaulieu, and Sankyo cameras. ∎

NOTE: I have done my best to give the reader a sense of the lifespan and cost of Super 8 equipment by including the years of manufacture and original retail price of equipment in the US. Please note that I came across much contradictory information in this regard. Further complicating issues is that equipment was often previewed at trade shows in a certain year, but then released to the public at a later date. Dates of manufacture for projectors were more difficult to track down. The dates listed tend to reflect the original year of manufacture rather than the projector's lifespan. Information for editing equipment proved even more difficult. As this book is not meant to be a comprehensive catalog of Super 8 equipment, please forgive any inaccuracies. The goal is to give the reader a sense of the equipment landscape at various points along the Super 8 timeline.

Beaulieu 4008 ZM II, 1971–1976, $899

Minolta Autopak-8 D 10, 1969–1973, $580

AMP Sound Super 8 Model 2, 1970–1971

Kodak M2, 1965–1968, $46.50

Canon Auto Zoom 1014, 1973–1979, $830

Bell & Howell Filmosonic 1227 XL, 1977, $489.95

Nizo Integral 7, 1979–1985, $1,003

Bolex 280 Macrozoom, 1971–1973, $319.50

Agfa Microflex Sensor, 1969–1970, $199.95

Yashica Super 600 Electro, 1971–1972, $129.50

Minolta XL 660 Sound, 1976–1978, $400

Fujica Single 8 Z600, 1968–1972, $329.50

Minolta Autopak-8 D6, 1970–1974, $234.95

Bauer C14 XL, 1978, $200

Elmo 1012S-XL Macro, 1978–1981, $999.95

Nikon R 10, 1973–1979, $950

Technicolor Mark Ten, 1967

Agfa Movexoom 3000, 1970–1975, $298

Zenit Quarz 1x8C-2, 1974–1983

EQUIPMENT

Specialty Cameras

NOTES ON SPECIALTY CAMERAS: The preceding discussion focuses on the typical range of Super 8 cameras found on the market. These are the cameras that accepted film cartridges, and if they were sound cameras, they were single system sound cameras. There were a handful of rare breed cameras that bear mentioning.

Double Super 8

Double Super 8 was a format that never entirely caught fire. Canon and Pathé were the only manufacturers in the pool on this one. They essentially modified 16mm cameras to work in the Super 8 realm. In a certain regard, they functioned similarly to Regular 8. In other words, the film you got from Kodak came in 100 foot rolls that were 16mm wide. When you shot, you only exposed 8mm of the film. When you got to the end of your 100-foot spool, you flipped the spool over, reloaded it, and shot for another 100 feet, this time exposing the other 8mm of film. The lab would slit it, splice it, and you'd end up with 200 feet of Super 8 film. The Pathé could even accept 400-foot loads of film. On the plus side, you could shoot well beyond the 3 minutes and 20 seconds offered by conventional Super 8 filmmaking. The downside was that the hefty footprint of the camera negated the unobtrusive nature of more conventional Super 8 filmmaking. The Pathé also came in at a whopping $2,995. The Canon was a steal at $1,300.

Single 8

Fuji chose to mark their own territory in the realm of 8mm filmmaking. Their Single 8 format was essentially the same as Super 8, but the cartridge design was different. You couldn't load a Single 8 cartridge into a Super 8 camera, and likewise, you couldn't load a Super 8 cartridge into a Single 8 camera. At a technical level, in the Kodak cartridge design, the pressure plate existed inside the Kodak cartridge. The pressure plate is an essential element in a camera system that ensures your film will be in focus. In the Fuji universe, the pressure plate was not part of the cartridge system, but part of the camera. But once you got the film back from the lab, you could play a Single 8 film on a Super 8 projector and edit Single 8 with most Super 8 editing setups because the perforation size, shape, and distance between perfs was the same in both formats. Single 8 had a polyester base as opposed to Super 8's acetate base, which led to a thinner, but stronger film. Due to the stronger base of Single 8, you couldn't cement splice it, so all editing of Single 8 lived in the realm of tape splices.

Double System Sound

One of the beauties of Super 8 was the immediacy that came from working with single system sound. You could shoot a film, and as soon as it came back from the lab, you could be editing on a top-of-the-line edit system that was about the size of a toaster.

There were some folks, however, who wanted to take Super 8 into a more professional realm. Designing a double system setup was the way to do this. One would need a camera running in perfect sync with an audio recorder. This was referred to as double system because one system recorded your film while a second system recorded your audio. There were several double system setups on the market. Bell & Howell was the first in the pool with the release of their Filmosound 8 in 1968. However, the Bell & Howell system was not reliable and was pulled from the market in short shrift. The system Richard Leacock designed at MIT used a Sony cassette recorder to capture sound. At Harvard, Bob Doyle modified a Sony reel-to-reel quarter-inch machine to record on Super 8 fullcoat (a fully coated magnetic film that was 8mm wide and sprocketed for 8mm use). For 16mm productions, 16mm fullcoat was a standard tool of the trade.

After shooting, you would edit your film on a flatbed editing system or an editing bench with a sync block to run your picture and your audio simultaneously. When you completed your edit you would still have separate reels of audio and picture. You would then add a sound stripe to your film and transfer your audio to the newly minted sound stripe on the film.

That's a quick glance at the whole process, but which cameras would work for double system? It's complicated. Some cameras came equipped with single frame switches that could send out a sync pulse to your audio recorder. Some cameras were designed with the ability to send out a pilotone sync pulse to your recorder. Some equipment utilized highly accurate crystals to ensure that both camera and sound recorder ran at the same speed. This process was called "crystal sync," which alleviated the need to connect the two devices with a physical tether. Regardless, the ultimate goal was to send a pulse between the camera and recorder that would allow them to maintain sync over the course of shooting.

Underwater Photography

Believe it or not, you could shoot Super 8 underwater. Reasonably priced housings were made to accommodate Kodak, Fuji, Beaulieu, Bell & Howell, and Eumig cameras, to name a few. For under $200, you could begin to explore the icy depths of your favorite body of water in Super 8. ∎

EQUIPMENT

Fujica Single 8 P400, 1972–1977

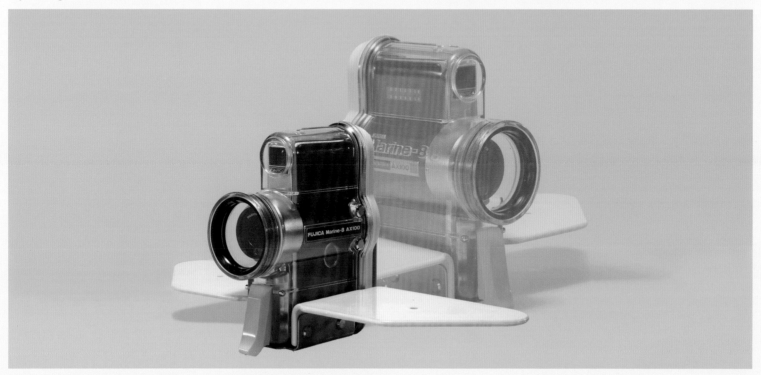

Fujica AX100, 1973–1978, $124.95. Marine-8 Underwater Housing, $89.95.

ESSAY

My First Camera

TO SAY I WAS A SELF-TAUGHT filmmaker is a bit of a lie. I did take one Super 8 class and one 3/4" video production class in college. I have little recollection of the topics covered or what cameras we used in those classes. In my Super 8 class I made a pretentious film about how college libraries were soul-sucking. The film was black and white, somewhat stultifying, and definitely silent. This leads me to believe that we only had access to silent cameras. But a silent world was not for me. I needed sound. How else would my characters project their potty-mouthed missives?

It was 1985, so finding a new camera was a challenge. Part of the problem was that I didn't even know where to look for a new camera. Very few stores were still carrying Super 8 cameras, but given that Super 8 was a home-movie medium, it was easier to ask around to see who already had one. I found my guy: Mike Woolson. He was great. He had a shock of red hair, was a budding young cartoonist, had a worldview influenced by *MAD* magazine, and had an unhealthy fixation on all things Adam West-era *Batman*. It helped that we were friends and worked together on an underground, *National Lampoon*-inspired college humor magazine. Relevant to me was that Mike had a camera and he wasn't using it all that much.

Was it a good camera? Was it a bad camera? I had no clue, but it was a camera and Mike was willing to lend it to me, so it pretty much checked off all my boxes. As it turned out, his GAF 805 M was pretty sleek. Most Super 8 enthusiasts never seemed that hot on the GAF cameras. They didn't have the caché of the more highly sought after brands, but I loved this camera. It had a killer lens. It sported a 7.5mm to 60mm zoom lens, which was a big zoom range for Super 8, and 60mm was long for the Super 8 world. It was a sound camera and it was solid. This GAF was a hefty brick. Perhaps best of all, it looked pretty darn cool.

The folks at the General Aniline and Film Corporation (GAF) clearly were willing to think outside the box when it came to industrial design. Check out the handle design on the 805 M. This is not how the handles on most Super 8 cameras looked. Most cameras had a rectangular or slightly rounded handle that ran perpendicular or at a slight angle to the camera's base. All told, most cameras exhibited some pretty meat-and-potatoes industrial design when it came to handles…at least compared to GAF models. Most of the cameras pictured in this book will attest to that.

And let me tell you, people were afraid of the GAF handles. Whenever I shot with my 805 M, I got quizzical glances. It was viewed with distrust. But I loved it. It had a good weight and felt properly balanced. I had no complaints.

Between 1985–1987, I shot four of my early sound films with that camera. In 1987, it was time for me to pull up stakes in Ann Arbor and head west. The camera had been on permanent loan from Mike, but now I needed to turn that loan into a personal holding. I made him an offer. Eighty dollars. Sold. The camera was mine.

That camera served me faithfully. From 1987–1996, I made another seven films with it. When making my 1996 film, *I'm Not Fascinating,* the camera temporarily stopped working, forcing me to use some lesser models for the bulk of that film. Subsequently, I was able to revive the GAF, but its days were numbered as I was growing less confident in its reliability. As luck would have it, a better model came along. My new permanent loan was the highly-sought-after Canon 1014 XL-S. Not much could compare to the 1014 XL-S, and my beloved GAF was relegated to the back of the closet.

Mechanics aside, the GAF's body began to give out. Its plastic, metal, and rubber parts seemed to be corroding in front of my eyes. It was as if the camera had leprosy. Every time I picked it up, I was left coated in a black, rubbery ooze that was really hard to wash off. For the most part, I couldn't even figure out where this glop was coming from. The handle? The body? It didn't make any sense. What I could ascertain was that the eyecup was clearly a culprit. After shooting, I looked like Spot from *The Little Rascals* with a big black circle tattooed around my eye.

Alas, the GAF had a good run. It now sits in a cupboard in the basement, entombed in bubble wrap. RIP. ∎

"Was it a good camera? Was it a bad camera? I had no clue, but it was a camera and Mike was willing to lend it to me."

ESSAY

GAF 805 M, 1976–1977, $465

INTERVIEW

Bob Doyle: Bringing Double System to the Super 8 World

BOB DOYLE WAS ONE of the pioneering architects of double system sound, which led him to start Super8 Sound in Boston in 1972. I talked with him about the origins of the idea and the research and development necessary to pull off this technical feat. Super8 Sound was at the forefront of professionalizing Super 8 production. Doyle left the company in 1977. Super8 Sound moved to California in 1987 and morphed into Pro8mm in 1998.

What was your interest in Super 8 before you developed double system and started Super8 Sound?
I was an assistant director of my Harvard College Observatory in the department of astronomy. At Harvard I was a visiting member of one of the Harvard houses. We were all expected to support students and things they wanted to do. A junior at Harvard, Robert Cole, wanted to make a movie for an astronomy class instead of writing a paper. The professor in the course asked him to work with me. I said, "Okay, I'll build you a system." I had been making home movies with the Bell & Howell Filmosound, which predated any kind of a double system. It had a tape recorder and a projector. You could record with a camera and the tape recorder, and then you would play it back with their projector. It seemed nice, but it was clumsy and it really wasn't full double system.

I learned Ricky Leacock at MIT had received a huge grant of 300,000 dollars from the founder of Polaroid, Edwin Land, to make a Super 8 equivalent to the 16mm tools that he had designed and built with sync sound on a separate recorder, both a cassette recorder and a kindred tape recorder, which could record on magnetic fullcoat film. That was so obvious to me, to move the professional methods that 16 and 35 were using and getting them down to Super 8. I immediately thought that I could build something like that. I thought I could take a tape recorder and convert it from 1/4", which is something like 7.35mm to 8mm by enlarging the guideposts on the recorder. All I thought to do is to unscrew the little guideposts and make new ones myself by grinding them off on a lathe. I had a friend who did that sort of work, and he and I and my brother-in-law all decided we could adapt a Sony tape recorder. And some funny things about this particular recorder, the TC-800B, it belonged to Bob Cole. He said, "Okay, I'll contribute my recorder and see if we can turn it into our fullcoat recorder."

So, I designed a circuit which would do two things: it would take a signal from the camera if the camera had a one-pulse-per-frame switch built into it. It closed the switch each time it took a frame. I then designed and built a circuit in my own oven in my house. You cook it and certain tracks that you draw on it get to be conductive and other ones are not. I made the circuit board, and put it in the recorder. We started in October 1972 and by November 1972 we had a working circuit that seemed to be synchronizing. We ran a full page ad in *American Cinematographer* in April of 1973. We told people we could sell them a 500-dollar recorder. The ad claimed to be able to not only record your sound directly on fullcoat, we thought you could also record directly on a cassette tape like the Filmosound system, and then you could "resolve" the sound and get it converted onto magnetic film fully coated.

Let me back up. Bob Cole wanted something to make a film. I started researching and I found that Ricky Leacock had his first prototypes for a system that would sell for $7,000. It was like the cost of a Volkswagen in those days, and he'd say to people, "Look, you can either have

"That was so obvious to me, to move the professional methods that 16 and 35 were using and getting them down to Super 8."

INTERVIEW

a car or you can be a filmmaker." So I applied to Harvard's president, because I was in one of the houses, and the houses had the ability to ask the president for support. I wrote a proposal and asked for $7,000. But he said, "If we give your house $7,000, [the other houses] are all going to want $7,000 for something similar. We can't do that just for your house." So I said, "Okay, I'm going to just build it myself." So Ricky Leacock had a system out and being tested while I was building one. Then Ricky sold a whole bunch of systems all over the world. The government of India bought five of them. Ricky went to India to meet with Indira Gandhi, and it was a whole idea that we were going to lower the cost of filmmaking. It was a very exciting time. And one of the things that happened is that the MIT-Leacock systems started shipping, and immediately there came back a lot of problems. It lost synchronization. It took them a while to figure out what was going on. It turned out once you spliced the magnetic film, cut it into scenes and spliced it with splicing tape, it accidentally counted the splice if the splice widened open a little bit. It would think another frame had gone by. It turned out that the servo control loop by which it counted the frames in picture and sound would add another frame every time there was a splice that [stretched]. I had written a system where the servo loop was very gentle and it assumed that if you cut two counts within 1/18 or 1/24 of a second, it would neglect it. So my system worked, and his didn't. The next thing you know, we're starting to sell Super8 Sound recorders to everyone who bought an MIT-Leacock system. They then paid another $500 to get a recorder that worked.

Your system was so much cheaper.
Oh, yeah, but my system was not everything theirs was. Theirs came with a camera, it came with both a cassette recorder and a

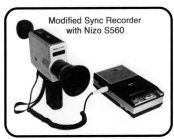

fullcoat recorder. It came with an editing table, a kind of flatbed like a Steenbeck. So that was already $2,000 or $3,000. Ours was just the recorder, but we promised that we would have mechanical synchronizers.

The sad part of Ricky Leacock's project was that it partly didn't work, and ours was so much cheaper, that they went out of business. And I hired the two young guys who had left MIT and gone to the company Hamton Engineering, which was making the Leacock system commercially.

So it was an effort to me to get into business and see if I could run a business. The cost is $125 to make it, and we'd sell them for $500. If we could make one of these a weekend and sell them, that would be nice extra income. I thought we'd make fifty a year. It turned out that over the next three years we sold 2,000 systems all over the world. By '76, we were doing three quarters of a million dollars of business or thereabouts.

Who are you selling all those systems to? Was it schools, or documentary filmmakers, or news stations?
I had a variety of customers. State University of New York bought five systems. Many individual schools bought them. I gave them to the students at the Harvard Radcliffe filmmaking workshop. We sold to individuals, lots of people who became major filmmakers.

Before yours and Leacock's system, there was no way to do double system with Super 8 other than the Filmosound, correct?
Well, there were a couple of things that looked like they had a big moment. There was Optasound, and there was the Cine Slave System, but they were kind of expensive compared to what I did. Cine Slave worked with various recorders. I don't think Cine Slave continued for many years. Optasound came up for quite a while.

INTERVIEW

Did they predate you?
I heard about them at the time I was building my Super 8 sound recorder. They used just a 1/4" tape with little holes in it. They built a machine that would move them together—the picture film and their funny little perforated audio tape. I always tried to write about Optisound in things I wrote about in those days in my Super8 Sound catalog. We had a 15,000 circulation in our best year. A lot of people said to me they learned everything they needed to know about making filmmaking because instead of just my recorder, I promoted all the things you needed. When I learned something, I would write it into this seventy-page catalog and we printed quite a few of them.

I started making films in the mid-eighties and I didn't see any of the double system equipment. They'd sort of disappeared. Do you have a sense of why that is? Was it the introduction of video?
I basically made the argument that Super 8's resolution was getting better all the time. They were always better than electronic systems like video. But it looked like there would become a day when you could record a picture in the Super 8 and release your work in video. So that was my concept of Super 8 video. A lot of people said, "Hey, Bob, you're abandoning film if you go to video." And I did go to video, big time, but I always said that Super 8 allowed you to get an image that had a film look. I don't know if Roland [Zavada] mentioned the importance of the Kodak Videoplayer? That was the thing that kept me and my catalog and sales going. It was a flying spot scanner, so it had the ability to take work and turn it into editable video. Kodak said to me, "Bob, we're going to let you be the first company to sell it. We're going to let you do that because you have this international reach." By 1976, we had deals in India, Canada, France, Germany. We were known as the one that was professionalizing Super 8. So I got to introduce the video player, and that fit with my view that double system sound was not necessarily going to be the way it was going to wind up. It was going to get converted into video as the delivery medium. ∎

Tommy Madsen: Super 8 Camera Design in the New Millennium

IN 2013, A NEW SUPER 8 CAMERA hit the market. Produced in Denmark by Logmar Camera Solutions and designed by a father and son team, the emergence of the Logmar Super 8 camera got people fired up. After all, this was the first new Super 8 camera designed in almost thirty years. The camera was an extremely limited-edition release and quite expensive, but it inspired Kodak to ponder introducing their own new Super 8 camera for wider release. I talked with Logmar CEO Tommy Madsen, who designed the Logmar Super 8, about how their camera came into being and his involvement in the development of the new Kodak camera.

Give me a little background on the Logmar endeavor.
We made our first Super 8 camera in 2013. That was a pin-registered Super 8 camera where you made a loop at the actual cassette and ran it through a metal gate where you have a pin that was engaged into the perforation. You cannot say it is a normal Super 8 camera. It was more difficult to load. But the results from that camera, I think we can say, are the best Super 8 ever.

That's a different design than original Super 8?
The original you just [put] in the cartridge and pull the trigger and that's it.

What made you want to make a new Super 8 camera in 2013?
I've always been interested in filmmaking. I was born in 1950 and then in 1960 I

INTERVIEW

Tommy Madsen (left) shoots 65mm.

bought my first Regular 8 camera and in '72 I bought my first Super 8 camera. In the eighties they stopped building Super 8 cameras. I think the last camera was in 1982 and then came all the video. I never had a video camera. I only had my film camera. In '95 I got my first computer and I realized on eBay there was a huge amount of old cameras for sale. I saw all these models we dreamed of when we were young, but we didn't have the money because they were expensive. I started buying some of the more expensive models to test out. The problem with buying a camera on eBay is that normally you have to buy several to get one that is actually working. The mechanics are nearly always working in these cameras, but electronics from the sixties and seventies don't normally work today because the electrolytic capacitors are dried out and all sorts of weird things

happen. Your light meter doesn't work, and maybe the motor can run but it's uneven in speed. From 2010 to 2013, I always had this dream of trying to build my own Super 8 camera. My education is electrical engineer, and my son has a degree in electronics. I asked my son could we make a Super 8 camera where I made the mechanics and he could make the electronics? He actually stated, "Who the hell would buy a film camera in 2013?" I laughed. But from '95 until that time, I realized that there were two huge forums on the internet, an American one and a Norwegian one. They actually discussed a lot about a new Super 8 camera and it never happened. That was one of the reasons that I thought we should try doing one.

Given the Super 8 connection to home movies, I think there's something wonderful about the new camera being built by a father-son team.
I cannot give you the name and workplace for my son because he has an NDA agreement. His work would not allow his name around the world. We managed to make a prototype and there was a lot of 3-D printed parts as well as metal parts. We actually managed to make very good prototypes where we had some really good results on the footage. When we had built that camera, my son asked, "So, what now? What are we gonna do?" I said, "Good question." He actually said, "I'm putting a photo of it in one of the forums." I said, "Do you think that's a good idea because we don't have a real camera yet?" He said, "Let's try it." He teased it a bit on cinematography.com. He said, "What if there was a Super 8 camera who could do that and that and was made like this and this?" A lot of people said, "We've heard that before. That's an old idea and it'll never happen." Two hours later he put up the photos and said, "It's actually here. It's the first prototype." And then the whole world got ballistic. That's how it started. We knew that there was still a market in the US and we were aware of Pro8mm in LA. We contacted Phil [Vigeant] who was the chief of the company to see if he was interested in bringing it to the markets. When we went there, we showed it to some of his customers. From there we raised the money to build fifty cameras. In 2013 and '14, we managed to sell forty cameras. I think the reason why we only managed to sell so few was because it was quite an expensive camera. It was by no means a cheap consumer camera. When the forty were sold, nothing more happened for nearly a year. Then, Phil contacted us in the beginning of '15 and asked if we would like to participate with him in Cine Gear, which is a huge exhibition in LA. We told Phil that we couldn't afford to participate because we hadn't made any profit on the

INTERVIEW/ESSAY

"He actually stated, 'Who the hell would buy a film camera in 2013?'"

initial sales and quite frankly were thinking about obsoleting the camera due to lack of sales. Phil was quite clever, however, and asked Kodak if he could participate in their booth and present his old and refurbished cameras. Kodak said yes, and instead he took our camera with him and put it on a tripod. I remember one guy from Kodak said they couldn't understand why there were a lot of people at Phil's Super 8 exhibition. Sometime after the fair Chris Richter, who was working for Kodak, called us to inquire if we could help them build a new Super 8 camera. We weren't really sure what to expect, to be honest, but he persisted and invited us for a meeting. Two or three days later we were in the car driving to Hamburg. At that meeting I had some drawings with a more normal Super 8 camera I made a long time ago. After a few hours of conversation, he finally said, "Let's do it." This happened the 3rd of August 2015. We set out and created CAD drawings between then and January 2016, when Kodak announced the camera at the Consumer Electronics Show. Following that it was decided that we should build ten prototypes that should be able to shoot film and be tested. In July the same year, the ten prototypes were finished, and we went to Berlin to get them tested in a climate chamber where you can simulate, for example, high humidity. One of the tests was +fifty degrees centigrade and 95 percent humidity. That's the same as a tropical climate. And also in very cold climates and normal climates. All of the cameras survived with minor incidents and issues, which were later fixed once we came back to Denmark. In the fourth quarter of 2016, our prototypes got the final approvals from Kodak and we were feature complete.

What's your favorite part of the new camera?
In the old Super 8 cameras, in the film gate you have this plastic pressure plate in the cartridge. So, what we did on the Kodak camera was the gap between the film and the gate and the pressure plate in the cartridge, we have fine-tuned it. It's more narrow than in the old Super 8 camera, which gives a more crisp and sharp image. We also made it run perfect so you can record sound. The best part is the LCD instead of the optical viewfinder. If you look at old Super 8 cameras, if you look through an optical viewfinder and you have low light, you won't see a thing. It's much easier to see a good image on an LCD in low light because of today's technology with LCD screens. Those are the three things that we wanted to bring up on our modern Super 8 camera. We couldn't make a pin registration because then people would have a lot of trouble loading the film. So, it is based on you plug in the cartridge and then you pull the trigger.

Now you just have to make a new Super 8 projector and your job will be complete.
I actually would love to make a new projector. I also have some good ideas how to make a good projector, but I don't know if Kodak is interested in such a project.

They're bringing back Ektachrome, so they have the reversal films now.
And the real way to see that film is definitely on the wall. ∎

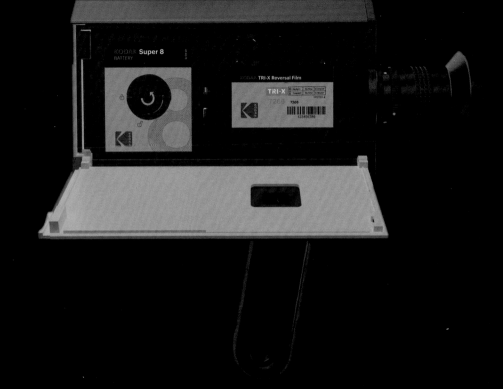

Prototype of the new Kodak Super 8 camera.

CHAPTER 4:

FILM STOCKS

FILM STOCKS

To anyone who has grown up in the era of digital video, the importance of choosing the right film stock can't be overstated.

Film stocks were an essential factor in determining the look of your film. Today, when we shoot video, the camera's sensor is the major factor influencing the look of your video. But once you've got your camera, the brand of CF or SD card you use doesn't affect your image quality in the slightest. The look of your video is tied to your camera and accessories like lenses. While those elements certainly were a factor, you also had a variety of film stocks to choose from, which were instrumental in shaping the look of your film.

When the Super 8 movie system hit the streets in 1965, it was released in conjunction with Kodachrome II, a beautiful, fine-grained, color film stock. Before long, a number of other stocks were available for purchase, not just from Kodak, but from companies like Agfa, Fuji, GAF, and 3M. There were color film stocks, there were black-and-white film stocks. There were stocks that had a lot of grain and there were stocks that featured a fine grain. There were high-contrast stocks, there were stocks that minimized contrast. There were stocks that oversaturated the color and there were stocks with a more natural palette. Film stocks produced by different companies would interpret color differently. In other words, Kodak film stocks looked different from Fuji film. According to Roland Zavada, who was on the team that designed Kodachrome II in the 1960s, Fuji, for legal reasons, had to avoid directly copying how Kodak made their stocks, which led to a "difference in the photo interpretation of the dyes used in Kodachrome versus those used in Fuji color."

Super 8 didn't offer quite the array of film stocks compared to what was offered for 16mm and 35mm, and this was one of the format's limitations. That said, there was a range of available stocks running from good to "fugly." It should be noted that many film stocks came and went through the years. For historical reference I will focus on the five Kodak stocks left standing by the mid-1980s, as well as explore the remaining options available today. Perhaps this is an arbitrary marker, but 1988 represents peak film stock in terms of "the number of different films that were made" across all Kodak film and photo formats. Additionally, much of the Super 8 work that you can track down today was made utilizing one of the stocks listed below.

You could purchase the color film stocks as either sound or silent film. The black-and-white film stocks, unfortunately, came only as silent. If you wanted to make a sync sound black-and-white film, you were out of luck, unless you wanted to get inventive, or had the ability to shoot double system sound.

Kodachrome 40
1974–2005

Arguably, Kodachrome 40 was one of the most beautiful film stocks on the market. The color palette was stunning. It was rich and saturated. The reds and the blues were absolutely vibrant. I always described the colors as "hyper real." If you wanted to film a circus, this was the stock for you. Talking about color with Roland Zavada, he notes, "There's always a question of what is the color of blue sky. Is it Kodachrome blue? Because Kodachrome reproduced the blue skies with significant emphasis. I would call it a fault because it did not faithfully reproduce nature. It enhanced certain characteristics." While the engineer may not have been pleased, this enhancement was one of the factors that made Kodachrome such a beloved film stock.

Kodachrome was considered a slow film (not very light sensitive), with a low ISO/ASA/EI of 40. This meant that you needed a lot of light to get a good exposure, and as a result K40 was considered the outdoor stock for Super 8 shooting.

If you shot on a bright sunny day, the colors leapt off the screen. Kodachrome was a favored stock, not only of Super 8 filmmakers, but of 16mm filmmakers, and still photographers. One of the factors that gave Kodachrome 40 its rich color palette was its complex dye structure which necessitated a seventeen-step processing procedure. Not all film labs could process it, and for best results, you needed a lab that specialized in Kodachrome processing. By the late 1980s, Kodak operated "ten labs that processed Kodachrome," with another "ten non-Kodak owned labs" processing the stock as well.

Manufacturing of Kodachrome came to a halt in 2005. While the ferocity of the run on photo shops to snap up the remaining inventory of K40 wasn't as severe as the run on US banks in 1929, Kodak's decision to discontinue the beloved Kodachrome sent a shudder through the film community. Kodak continued to process the stock at their plant in Lausanne, Switzerland, until September of 2006, after which Dwayne's Photo Lab in Parsons, Kansas, took up the processing mantle, with the final rolls of Kodachrome being delivered to Dwayne's for processing on December 30, 2010. While a T-shirt printed by Dwayne's commemorates the December 30 date, actual processing eked into the New Year.

Kodachrome can still be cross-processed as black-and-white film, but that's a whole other can of worms, and the purist in me thinks, "What's the point?" On the other hand, the artist in me thinks, "What the hell!"

Ektachrome
Ektachrome 160 Type A, 1971–1996
Ektachrome 160 Type G, 1974–1996
Ektachrome 125 (VNF 7240) 1997–2004

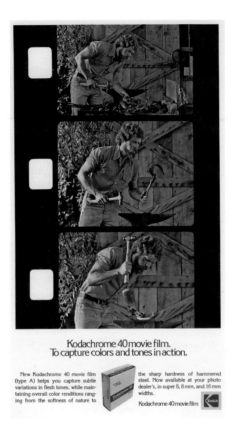

Ektachrome 64T, 2005–2010
Ektachrome 100D, 2010–2012
Ektachrome 100D, 2018–present

The other readily available color film was Ektachrome. The chemistry of Ektachrome has changed over the years, and that's a good thing, because in the mid-1980s there was no doubt about it, Ektachrome 160 was plug ugly. Lodge browns and insane asylum greens were the color palette. And while that might sound cool, it really wasn't. Films shot with Ektachrome always felt a little sickly.

Ektachrome 160 came to market in conjunction with Kodak's XL camera line in 1971. This was their "existing light" camera, which offered the promise of being able to make movies in low light with no need for special lighting equipment. The ad campaign features a lot of people filming around campfires, in bowling alleys, and glamour shots of household lamps. While I might complain about the overall look of Ektachrome, the ability to shoot in low-light situations was a boon for Super 8 filmmakers. It really did make the filmmaking process that much easier. This was a major initiative that was spearheaded by research chemist Don Gorman.

Like most of the Super 8 film stocks, it was properly color balanced for tungsten light. In 1974, Gorman played around with the stock's chemistry and developed a new version of Ektachrome called Type G, differentiating it from 1971's Type A. Type G could be shot in either daylight or in tungsten light without worrying about your camera's filter settings. While this sounds awesome, I would argue that Type G looked bad in all lights. Because it could be shot in a variety of lighting scenarios without having to concern yourself with filters, it was somewhat idiot-proof. As a result, it was often favored by first-time filmmakers. While Type G was easier to deal with on a technical level, it did no one any aesthetic favors.

Unfortunately, this stock largely defined the Super 8 look of the 1980s and early 1990s. Though Kodachrome was more pleasing to the eye, Ektachrome was easier to shoot. As already indicated, Ektachrome was a faster film (more sensitive to light) than Kodachrome, with an ISO of 160. This meant you didn't need as much light to get proper exposure. If you were planning on shooting a color film indoors, in all likelihood you would choose Ektachrome, solely based on its light sensitivity. You could shoot Kodachrome indoors, but you would need a pretty professional light setup to do so. This challenge was too much of a technical and budgetary leap for many starving artists and first-time filmmakers who usually opted for the most expedient, least expensive option, which meant documenting the indoor world with Ektachrome. If Super 8 gets a bad rap for being murky and ugly, this is why. Sigh.

1996 was a grim year on the Super 8 timeline. Kodak discontinued Ektachrome Type A & G, leaving Kodachrome as the only color film on the market. Though I've been bad mouthing Ektachrome, having a faster film in one's arsenal was an important tool for many filmmakers.

To many people's surprise, Kodak released a new Ektachrome in 1997. This stock had a much more appealing color palette than its precursor. People were psyched. This stock was cut down from a 16mm stock called 7240 with an ISO of 125. Not as fast as the older Ektachrome, but still significantly more sensitive to light than Kodachrome.

In 2005, a new color film, Ektachrome 64T, emerged. Ostensibly 64T was seen as a replacement for the recently departed Kodachrome. This new Ektachrome was not quite as sumptuous as Kodachrome, but it had a vibrant color palette that was well received by the Super 8 community. In 2010, 64T gave way to yet a newer Ektachrome, 100D. Too good to be true, Ektachrome 100D was put to death just a few short years later in 2012. But for some reason, you can't keep Ektachrome down. In October of 2018, Kodak shocked the film community by bringing back Ektachrome 100D and selling out of its initial run. A brand-new reversal film stock in 2018 was unexpected to say the least. Not surprisingly, this iteration of 100D has been met with a tremendous amount of enthusiasm within the Super 8 film world.

4-X
1971–1990

I loved this stock. It was black and white with huge, bulbous grain. It was the fastest (most light sensitive) stock in the Super 8 stable with an ISO of 400. This meant that you could take it into punk rock shows, shoot the band, and the footage would look gloriously gritty. It was a rock and roll stock. Alas, it was discontinued in 1990, only a few short years after I discovered it.

Plus-X
1968–2010

Plus-X was considered the outdoor film in the black-and-white realm. It rocked a 50 ISO, so like Kodachrome it was tough to deal with indoors. But if you were outside, it looked beautiful. It was fine-grained film with deep blacks and beautiful contrast. Kodak discontinued this gem in 2010.

Tri-X
1968–Present

Tri-X remains the lone standing black-and-white Super 8 film. It's

a 200 ISO film, so it has some nice rich blacks and a nice amount of grain that you can see flitting around your frame. It's a versatile stock that you can shoot indoors with the help of not too many lights. You can also shoot it outdoors without too much fear of overexposure. It doesn't have quite the personality of 4-X, but it's the last stock standing, so you have to take your hat off to it. Like Ektachrome, you can hand process Tri-X, which has always been one of its appeals.

Negative vs. Reversal

Another consideration when choosing film stocks was the distinction between negative and reversal stocks. When you shot negative, the lab produced a negative and a work print. You would assemble your cut using your work print. You could mark it up and scratch it without worry, because once you had settled on a cut of your film, you would go back and cut your negative, and generate prints from there. This was standard operating procedure for 16mm and 35mm.

Reversal film was like old slide film. You only got back a positive version of your film. This meant you were cutting your camera original. This was a challenging proposition.

The Super 8 world was a reversal world. Until the early 1990s, negative stocks weren't even an option. You only had access to reversal film stocks. All the film stocks listed above were reversal and they were the standard Super 8 film stocks that you could use in most any camera on the market.

Reversal film stocks offered their own set of challenges. These stocks had less latitude than negative film stocks. Latitude refers to the amount you can under or overexpose your film and still get an acceptable image. Latitude varies stock to stock, but generally speaking you had significantly less latitude with reversal stocks. This meant that when you shot reversal film, you had a small margin for error in terms of your exposure readings. If your film was shot under or over, you had little hope of correcting this in post, unless you were paying for a high-end transfer. Super 8 stocks gave you about 1 1/2 stops latitude in either direction. That was a slim margin, so your goal in production was to get your exposure right the first time.

I mention this here because in 1993, Super 8 Sound, now called Pro8mm, began cutting down 35mm negative films stocks for Super 8 usage. This was a great benefit because it opened up the number of stocks filmmakers had to choose from, thus giving filmmakers access to more looks. Super8 Sound offered up negative stocks with ISOs of up to 500. You could now shoot in lower light situations. Negative stocks also had greater latitude. The stocks that Super8 Sound produced gave filmmakers up to 3 1/2 stops of latitude in either direction.

Though these new stocks represented a great opportunity, they also created an aesthetic shift. The negative film stocks were only guaranteed to work in a specific set of high-end cameras. With these stocks you wouldn't be cutting or finishing your project on film. The negative stocks had to be transferred to video, all postproduction would happen in the video universe, and all exhibition would follow suit. At the time these services were first being offered, there were still plenty of opportunities to exhibit films on Super 8 and a healthy number of filmmakers committed to shooting, editing, and exhibiting on Super 8. The underground film world had not yet crossed the digital divide. As tantalizing as these new stocks were, many no-budget filmmakers were not making that leap, in part because video projection was still a dicey proposition at lower-end venues, microcinemas, and cinematheques. In 1993, there was an excellent chance that Super 8 projection would look better than a video projection of the same project. Also, let's not forget that many Super 8 filmmakers shot in Super 8 because it was what they could afford. Moving into the negative realm and having to do post in video drove up the costs of production, making this type of workflow cost-prohibitive.

Super8 Sound was embraced by people with larger budgets like music video producers, ad agencies, and Hollywood filmmakers wanting to insert Super 8 into their 35mm features. Oliver Stone famously used this process for *Natural Born Killers*. The difference between the Super 8 and 35mm was palpable, enhancing the gritty vibe of the film.

As the 1990s morphed into the 2000s, video became a more standard form of exhibition. As video became easier and cheaper to edit at home, using the Super8 Sound/Pro8mm stocks had even greater appeal. In 2000, I shot a film using their 200 ISO negative stock. It was exciting to have the faster speed, the latitude, the grain structure, and the color palette heretofore not available to me.

Iconoclasts like Neil Young and Jim Jarmusch tucked into this mode of production. Jarmusch shot the 1997 Young concert film, *Year of the Horse*, on Super 8. At a screening at San Francisco's venerable Castro Theater, the crowd roared in approval as the title card "Proudly shot in Super 8" came on the screen. Of course, Jarmusch tidied things up with a forty-track sound mix, but I suppose that can't be held against him. Young was back at it in 2003, with the release of the film *Greendale*—all shot on Super 8 utilizing Pro8mm stocks. To see that on the big screen was a Super 8 treat.

Beginning in 2004, Kodak got in on the negative bandwagon, introducing Super 8 negative stocks that corresponded with their 16mm and 35mm brethren. Currently Kodak offers three negative stocks, ranging from 50-500 ISO.

Today, given the ease and convenience of digital editing, the dwindling support for Super 8 postproduction services, the dearth of remaining reversal stocks, and all of the opportunities to share and distribute films online, shooting negative and finishing digitally is a workflow that makes a lot of sense.

In Europe, Kahl, Wittner, and ADOX all sell Super 8 film stocks that they cut and perforate from larger gauge Agfa, Orwo, and Ferrania films. Though the landscape seems to constantly be shifting, there are a surprising number of stock options out there for Super 8 filmmakers to take advantage of in 2019. ∎

INTERVIEW

Phil Vigeant: Super 8 Innovation throughout the Decades

PHIL VIGEANT STARTED WORKING at Super8 Sound in Boston in 1977. By the early 1980s, the company was struggling financially, as interest in Super 8 was on the decline. Vigeant still believed that the company had a dedicated customer base and bought it in 1983. He opened a branch in California in 1987 and renamed it Pro8mm in 1998. Under Vigeant's leadership, Pro8mm has continued to push Super 8 into the professional realm. Among their innovations are cutting Super 8 film down from 35mm negative and creating widescreen Super 8. Vigeant has a unique perspective on the lifespan of Super 8, having been involved with the medium since its early days.

How did you get involved with Super 8?
I got a part-time job when I was in college working at Super8 Sound as their bookkeeper. I was studying accounting and they wanted a person to do data entry.

At some point you bought the company. To buy a company you have to have faith in the product. What were you seeing as the potential for Super 8?
I worked there in college [in] 1977. All kinds of things were happening and everyone was into Super 8. There's people doing commercial productions and television show pieces and art films and documentaries, and [there was] tremendous amounts of positive energy surrounding Super 8. When I graduated school, I left that job and I got a real accounting job for a couple of years. I wound up coming back because I got a phone call from my friends at Super8 that they were having financial problems. I went down to check it out. This was probably around 1980, and everything had changed. The entire market for Super 8 collapsed around 1979, 1980. What was this huge thriving thing just evaporated. Companies were filing for bankruptcy left and right. Almost everyone that was involved in Super 8 went bankrupt except for the companies that were little divisions of big companies. Kodak didn't go bankrupt, but their whole Super 8 division went kaput. All the other companies like Beaulieu, Eumig, Chinon, Bauer, and Nizo, and all the names you know, had to fold out of the Super 8 business because there was none.

VHS came out in the late seventies. People got enamored with the idea that they could just pop the tape underneath your television set and watch it. You could get two hours of viewable material out of a ten dollars tape as opposed to paying for the film and the processing and setting up a projector. I read somewhere that Super 8 sales dropped by 95 percent in 1980. Just an amazing shift in their economics.

I showed up [at Super8 Sound], and they were like "What do we do?" I took home the books and I came back, and I told them, "You guys are in tough shape here. You owe hundreds of thousands of dollars, and the revenue is kaput. There isn't much to hold on to here."

I didn't really love the work that I was doing in accounting and it was a lot more fun at Super8 Sound. I craved being involved in that kind of stuff. Eventually, they offered me a job at Super8 Sound to try to salvage what was going on. We wound up filing bankruptcy along with all the other companies around 1981. The first bankruptcy didn't solve the problems. When an industry collapses like that it takes a while for it to settle out. After about a year, they needed to file bankruptcy again, and they were getting tired of trying to run this thing. I was still enthusiastic about being involved. We wound up striking a deal where I wind up taking over what was left of the company.

Given that sales have dropped through the floor and the financial scenario isn't great, what were you thinking?
Even though this huge majority of customers had left and went to video, there were still people showing up at the door every day who wanted to use this medium to do creative stuff. They were the initial reason for Super8 Sound to exist in the first place. I think they got swept up in the whole "everyone can be a filmmaker" frenzy that was the late seventies. What was left after all those wannabes went over to the next flavor was the core person who really had a desire to work visually and see film as a craft. Those people weren't going

INTERVIEW

Super 8 film perforating machine.

What prompts you to move from Boston to Burbank?

The market started picking up again. All of a sudden, this guy's doing a project in film and this guy's doing a project in film and before you know it, there's this kind of resurgence out of the ashes. Around '85, the music video came out and represented a real need for this creative process. A lot of young filmmaker types were connecting with musicians to do this sort of commercial project. The volume of that started growing quickly and it was all being done out of the West Coast. We were shipping half a dozen cameras every weekend to LA to shoot music videos and then get the film all the way back to Boston. We realized, boy, if we can do this much work in LA from Boston, imagine when we could do if we were in LA.

What cameras were you shipping to people?

It was all the Nizo. The 6080 and customers were buying the Integral because there was a huge inventory of those cameras that somebody had warehoused in New Jersey. They picked up the inventory of Nizo, and they couldn't find anybody to buy these things, and we had a way of selling it. They were very aggressive about getting us to take more and more of them. Dropping the price. We sold a 6080 for about $1,800 dollars to begin with. Every time we'd sell three of them they'd say, "How about if you buy them for a $100 less?" We were [eventually] selling 6080s for about $1,000. I think they had over 500 cameras at the time, and we wound up selling every one of them.

How did you come up with that idea to perf 35mm for Super 8? At that point Super 8 is all reversal stocks.

A bunch of things came together in the perfect storm. In '92, Kodak was discontinuing everything. They had sent out a letter: "We're going to discontinue Ektachrome." They were getting rid of the VNF, the A, the G. We're like, "We're screwed. We're not going to have any stock that we process. We're going to lose customer base." So we're hemming on what to do. Customers had always been jealous that they couldn't get the same stocks as they got in 16 in Super 8. This always screwed us up. If we got on a set, they'd like the look of the Super 8, but we'd be on a set that would be lit for a 200 tungsten negative film, and we'd be shooting Kodachrome and it wouldn't match up with the lighting. It made it difficult for professionals to use Super 8 in context with other things they were doing. So we started thinking, you know, everyone's kind of bugging us about having this negative stock. We used to make Super 8 fullcoat, which, remember, is the reason the company was there. We actually manufactured the mag with a subdivision company of Technicolor. That had gone away because we weren't making fullcoat recorders anymore. We didn't need fullcoat, but the machines turned 35mm fullcoat into Super 8 fullcoat.

Then in the Boston office, this guy shows up from the Soviet Union with a reloadable Super 8 cartridge. I guess in Russia, the way that they pursue Super 8,

anywhere. They had real reasons why, in their mind, film was a more creative way of doing what they were doing, and they were sticking with it.

I sized up that there was many of them who just kept coming. There's probably a whole heck of a lot more of them still around, just hiding because they're embarrassed that everybody else had gone to video and they didn't want to say the word "film" too loud, because the general public was like, "Oh, this is old. This has been replaced by great new technology. We don't need this old crap anymore." So I think they were really happy that a place like Super8 Sound was still hanging in.

INTERVIEW

they don't have the Kodak cartridge, they have their own cartridges. They have little clubs and they would buy film in bulk and load their own cartridges to save money. [They] had these reloadable Super 8 cassettes which they were getting rid of as well. You take all those things together and you start going, "Wow, if we got a hold of that equipment at Technicolor that was the perfer and the slitter, and we could get those cartridges from Russia to load the film, and we don't need our Ektachrome machine anymore, because Kodak's discontinuing it. So, we had something we could convert to process negative; everything's there." Eventually we acquired that perf and slit machine and we were able to cut and perf 35 into 8. Then we got those Russian cartridges and we loaded it into Super 8. And we converted our Ektachrome machine to process color negative and by August of '93, we started making the first Super 8 color negative.

At the time the fastest Super 8 would've been Tri-X with an ASA of 200. What was then the fastest negative that you were cutting?
500.

Now you had the flexibility of negative, but also more and faster stocks.
Yes. Being able to do things the way everybody else is doing them. Super 8 was kind of autonomous because it worked with the old technologies of Kodachrome film and projecting. We were forcing it to do things like music videos, but it wasn't quite there because we didn't really have the tool set that they had in 16mm. You now allowed the Super 8 user the ability to shoot the same stocks as 16.

I remember you guys renting out camera packages, primarily Beaulieus. What cameras were you recommending that people use with that new negative stock?
Initially the cartridges weren't notched for the ASA. We didn't put that into the cartridges because we were selling the Beaulieu cameras and they didn't need that.

Because you could set the ASA on the outside of the Beaulieu, right?
Yeah. And the neg film is a little bit… well, we say it's thicker but I don't think the thickness is really it. It takes a bit more to go through a camera than the regular stock. With the Beaulieu, we knew that it would pull the film through without a problem. Some of the other cameras would have problems with the film jamming, so we wanted to keep the negative with the Beaulieu because we knew that it would work correctly.

You didn't want to go out and test all 300 models of Super 8 cameras that were out there?
It's actually a thousand different models out there. To have said to anybody anywhere that you could use this stuff was beyond what we were willing to say. Subsequently, that's what happened. People tried it in every cheapie camera. And as we got the feedback—"yeah, this stuff works fine"—we started providing little things to tell you how to cut out the notch. And eventually we started notching them ourselves, because it was like, hey, we have this thing that works great in your Sankyo or your Chinon.

And Kodak starts cutting down their own stock in 2004. Is that in response to what you're doing?
I don't really know for sure. We've always been pushing them to see the value of Super 8 in terms of the intro to film. If the first step that anybody takes in terms of making film is 16mm, you've got to provide a whole college education before you can even get something to look at. Super 8 was designed so that [you] could grab this thing and go make some pictures this afternoon, without a class. We've always been harping on Kodak to never get rid of Super 8 because this is how you get 35mm customers. They started to realize that if we want to continue to have customers we're going to have to support them. We're at first a little perturbed at them copying us, but then realizing that we were the largest motion picture lab that processes Super 8, that every roll they sold we had a pretty good chance was going to come to us for processing and scanning, and we made a

---

"If the first step that anybody takes in terms of making film is 16mm, you've got to provide a whole college education before you can even get something to look at. Super 8 was designed so that you could grab this thing and go make some pictures this afternoon, without a class."

lot more money processing and scanning than we did making film.

Another thing you brought was widescreen Super 8.
Once HD seemed like an inevitability, boy, we're in a square world and everything is going to be rectangle. Sometimes in the feature movies we've done, the DPs have wanted the Super 8 to be rectangular. We played a lot with anamorphic lenses to try to achieve this, but it's very cumbersome to seat an anamorphic lens in front of a Super 8 camera. We started working toward something that would allow filmmakers to get a rectangular Super 8. We didn't use sound anymore, so we could cut the gate out farther and make a wider Super 8 gate. We call it Max 8. It maximizes how much of the 8mm film is actually exposed to picture. Then we masked the viewfinder in the camera so that when you look through the camera you know where the picture is going to line up for 16:9, because it's still the 4:3 viewfinder.

Vigeant introduces color negative film stocks, 1993.

What have been the biggest challenges for you over the years in terms of keeping Super 8 alive?
The ball definitely keeps moving around. We went through a whole thing about five years ago with Urban Outfitters selling Super 8 cameras with rolls of film.

Were they selling the Rhonda camera?
Yeah. I was hanging out with the guy who was doing the Polaroid stuff. I got to know the US rep for the Impossible Project. He opened me up to this whole analog resurgence with Polaroids. And I'm like, "Who the hell wants to take a stupid Polaroid?"

Says the guy keeping Super 8 alive for the last forty years…
Well, you think you know what you're doing, but you probably don't! He opened me up to, "Oh, I guess there's a different way of looking at this whole thing." People want to do it because they could do it, not because it's sharper. They just want to be involved with the technology of old. It's cool to find an old thing and make it work. Before you know it, I'm talking to Urban Outfitters about selling cameras. We put together this thing called the film kit, where the film is sold with the processing and the transfer, and it's uploaded to you. We really put time into figuring out how we can make the logistics easy. Let's use the old mailer system. We'll sell the guy the film with the mailer, but go one step further. Let's actually deliver it to him over the internet so that he gets it right away. And then he gets the film later, so he still has the physical film. Then he buys a thing like a Rhonda camera, which is like there's no features really in it.

The Rhonda was a repurposed Canon?
It's the 310 XL, which has this phenomenal exposure range. It goes from f1.0 to f45. It literally takes a picture wherever you want. People are taking these things into nightclubs with 500 ASA film. We even took the names of some of these things off. We didn't use "50 Daylight." No. Sun film. Low-light film. No numbers.

It's like international signage.
Right. How simple could we make doing Super 8? Believe it or not, there's a huge group of people that started through that process. It gave us a new ground floor to bring people in. So you have these shifts that are going on. They're not always the same reasons why people are using this stuff. There's still a lot of people using it, and seems to be more and more. ■

CHAPTER 5:

PROJECTORS

PROJECTORS

Most gearheads in the film world obsess about cameras and lenses.

It makes sense to fetishize cameras. Film is a visual medium and cameras and lenses are prime factors in determining the quality of your image. But in the world of Super 8, I never worried too much about which camera I used. There were lots of good cameras with sharp lenses. It wasn't too hard to find a camera that would give you a nice image. A bad projector, however, could undermine or destroy a beautifully crafted image pretty quickly. And great projectors were a little harder to come by.

It was my belief that the projector was king. A great projector preserved the integrity of your film. A great projector allowed you to create a complex sound design for your film. A great projector determined the size of the venue in which you could screen your film.

As mentioned in the previous chapter, when you shot Super 8, you likely shot reversal film. When you shot 16mm, you likely shot negative film. This difference had major ramifications in the postproduction and exhibition process.

In the analog era, you made all your 16mm edit decisions by cutting your work print. The negative was sealed in a can and put away until the final stages of postproduction. Once all your edit decisions were made and you achieved final cut, you would go back and cut your negative to conform to the decisions made on your work print. At that stage, you handled your negative with the utmost care. You wore cotton gloves and edited in a sealed-off room that guaranteed a dust-free environment. The negative never ran through any type of viewing system. You cut it based on a series of edge numbers that ran along the edge of both your negative and your work print. All of this ensured that your negative remained unscratched, unscathed, and in pristine condition.

When you finished cutting, you sent your negative to the lab and they generated release prints, then stored your original in an environmentally controlled vault. Once you started sending your film out into the world for screenings, your prints would get beat up due to dirty projectors. Dirt and dust in a projector scratched films, and over time your print quality would decline. No worries, because you could always generate new, clean prints from your negative. This was the 16mm process.

The reversal film world of Super 8 was a wholly different beast. When you got your film back from the lab, you received only a positive version of the film. There was no negative. You cut your camera original film. Not only would you likely use that camera original as your screening copy, but it was also the source for any print or video transfer you would ever generate. As a result, editing became a daunting task. You needed a clean editing surface. Any fingerprints or dirt picked up during the editing process would have to be assiduously cleaned. You needed viewers that you could trust not to damage your film. Any scratch to the film's emulsion would remain until the end of time. Any damage inflicted on the camera original could not be fixed. Invariably, some level of wear and tear would be introduced during the editing process. Before you ever screened your film, it was already visually compromised.

When you were done editing reversal film, you had two choices in front of you. You could get a print of your film, or you could throw caution to the wind and project your camera original. This was a difficult decision. Why wouldn't you generate prints and keep the edited camera original as pristine as possible, treating it like you would a negative?

Generating a quality print from a reversal film was a trying, and often unsatisfying, experience. By the mid-1980s there

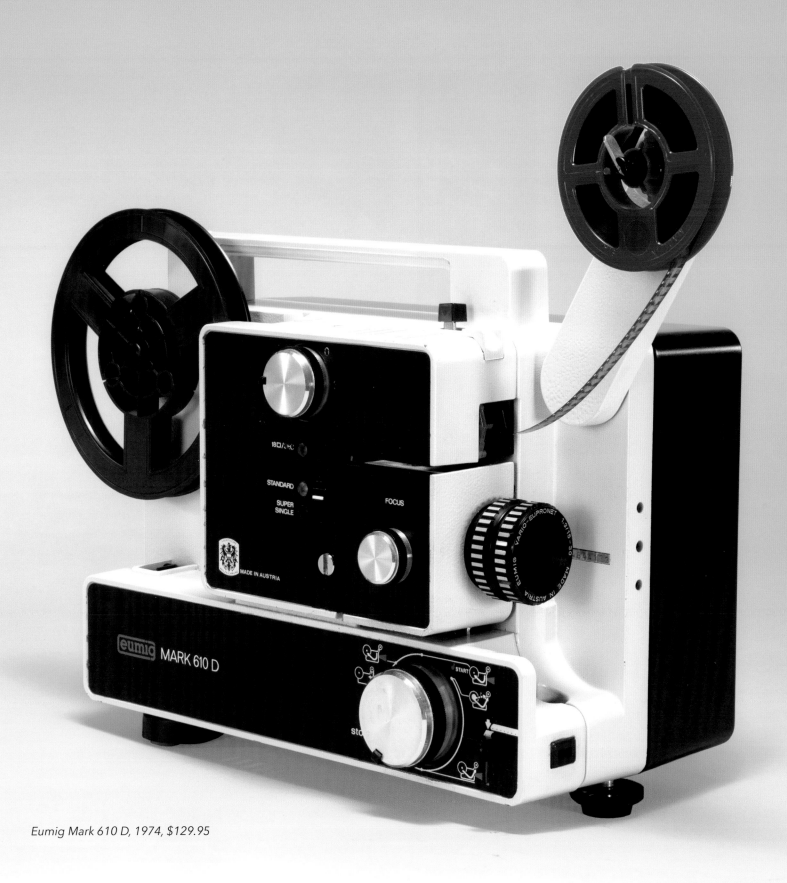

Eumig Mark 610 D, 1974, $129.95

were only a handful of labs making Super 8 prints. The quality, especially if sound was involved, could be dodgy. Also, any print from a positive film suffered generational loss. There was no way around it. You would lose image detail; shadows would become deeper, contrast would increase, colors would shift. Though you could get good prints, even the best of prints never looked as good as the camera original. Additionally, it often took several tries with a lab to get a decent print of your film, and there was a chance you would be paying for those attempts. At the end of the day, it was an expensive and frustrating experience.

Edge code numbers on 16mm negative and work print.

As a result, many filmmakers chose to show their camera original films. In many respects, it was easier and more satisfying to walk that tightrope. The literal rub, of course, was that each time you projected your film, you ran the risk of damaging your camera original. A nasty piece of dirt in a projector could cause scratches, etching a permanent line into your film. If the scratch occurred on the film's base, it would take the form of a black line. A colored scratch indicated you had damaged the film's emulsion. A white scratch meant the emulsion had been entirely scraped away.

Nowadays, when restoring a film, the black scratches can often be dealt with since they are only on the base, causing a refraction of the light. A process known as "wet-gate" scanning or printing fills in the scratch with a chemical, allowing the light to pass through the film normally. One is not so lucky with emulsion scratches since that is where the image lives, and once scratched away it cannot be replaced, forever leaving a clear spot on the film.

With this kind of risk, your projector became your best friend. A good, clean projector that was gentle on your film helped guarantee the visual integrity of your work.

Maintaining and monitoring the projector's cleanliness was paramount. I was fastidious in this regard. I kept my projector in a plastic bag, set inside a hard case, nestled in a very clean closet. On the inside flap of the case, I kept a cleaning log, charting cleaning dates, and logging the lengths of films that I projected. Every three or four hours of projection time would necessitate another cleaning. I was obsessed. Most people used rubbing alcohol to clean their projectors. I once got a tip from Super 8 gurus Bob Brodsky and Toni Treadway that Lemon Pledge was the way to go, and that became my cleaning agent of choice. I'd pull apart the projector, unscrew any moving parts, and clean the entire film path and miscellaneous parts with a Q-tip dipped in Lemon Pledge. If the Q-tip head emerged dirty, that section of the film path would get another pass until it was clean. The only part of the path that didn't get the Lemon Pledge treatment was the sound head. This I'd clean with alcohol. I'd spend twenty to thirty minutes going through this ritual, dozens of Q-tips discarded in a heap on the floor by the end of the process.

• • •

My Chinon 330 MV projector, bought back at the start of my life as a filmmaker, was my most prized possession. I bought that model, not based on any research that I had done, but simply because it was the one the camera store in my hometown had in stock. That model of Chinon was not considered a top-of-the-line projector. It didn't have the greatest throw compared to other projectors, and the focus could be slightly gamey. As it turned out, I lucked into a projector that was easy on the films, guaranteeing me the

Kodak M95, 1966–1975, $209.50

Kodak Instamatic M109 K, 1970–1974, $169.50

Heurtier Stereo 42, 1973–1974, $999

Kodak Moviedeck 457, 1977–1981, $247.50

ability to show camera originals countless times without worry of scratches. I traveled the world with that projector. I was often met with strange stares from airport security, not quite sure how to inspect a film projector, something they clearly never came in much contact with.

One of the exhibition/distribution limitations of choosing to work with camera originals was that you might turn down a screening if you didn't have faith in the projection. Sure, you could trust your own projector to handle your film properly, but trusting the projectors of total strangers was another thing altogether. If I traveled to a festival or screening, I would bring my projector even if I was told that the venue had an acceptable projector. I wasn't willing to leave things to chance. On more than one occasion I arrived at a venue only to find a projector that hadn't been cleaned in years. One memorable scenario was a screening held at a prestigious film society. In advance of the screening, I was repeatedly told how pristinely they kept their equipment. I was counting on this because it was a venue that my projector would not have a strong enough throw to work in. Upon arrival, I checked their projector and it was filthy beyond compare. Apparently, they had held a hand-processing and direct animation workshop the previous week. This meant that films with sharpie, nail polish, paints, dyes, wax pencil markings, glitter, and a variety of inks had been running through their projector. It took me well over an hour and hundreds of Q-tips to clean that projector.

The projection conundrum was real. I'd often get screening opportunities I couldn't attend. Would I send a camera original? First off, my film might get lost in the mail. Secondly, it might get eaten by a projector. I based decisions on venue reputations, and occasionally sent an original. However, in most cases I would decline the screening unless the opportunity seemed too good to pass up. It just didn't seem worth the risk.

• • •

The other type of projection challenge had to do with theater size. The bigger the theater, the greater wattage your projector's bulb needed to be. The greater the wattage, the greater distance the projector could be set from the screen. Additionally, there are different types of projector bulbs. Xenon-Arc bulbs produce a brighter, whiter light than their halogen bulb counterparts. In traditional movie theaters, 35mm projectors are outfitted with Xenon-Arc bulbs. San Francisco's Castro Theater, which sports a balcony and seats 1,400, utilizes a Xenon-Arc bulb that kicks out upward of 2,000 watts to properly illuminate the screen. That's the power needed to project in a space of that size. The type of 16mm projectors found in classrooms and microcinemas, usually sported halogen bulbs in the 250-watt range. While clearly not as powerful as the Xenon-Arc bulbs, a 250-watt halogen can still project in a good-sized space.

My Chinon, along with almost every other Super 8 projector, was not outfitted with such high-wattage bulbs. It had a 100-watt bulb, which was fairly standard for Super 8 projectors. This wattage was best suited for home movie or very small venue projection. If I were in an eighty-seat theater, or a bar or a club where the projector could be set within forty feet of the screen, I could project and feel confident in the brightness of the image. Once I started moving into bigger

Handling 16mm with cotton gloves.

GAF 3000 S, 1974, $309.95

Elmo GS 1200, 1977, $1,390

theaters, with throws beyond forty feet, image brightness started dropping, and that became problematic.

Often I'd be invited to show out of town and I'd bring my projector. If the show was in a large venue and I set up in the projection booth, my projector simply wasn't bright enough to project a good image. In these scenarios, I took to setting up the projector in the middle of the audience. I actually liked this type of screening. I felt it broke down the barriers between artist and audience. Independent screenings were often comprised of other filmmakers and aspiring artists; having the guest of honor plop down in the middle of the house had an invigorating and egalitarian quality that spoke to the independent film ethos of the day.

Of course, setting up in the middle of the venue presented a whole host of other problems. The little five-watt speaker on my projector wouldn't be loud enough to project sound to an audience of a hundred people. This necessitated running a long stretch of audio cable from the projector to the projection booth, in order to patch into the house sound system. At some point, I frankensteined a 100-foot audio cable that would allow me to set my projector 100 feet away from a sound board. The output on the projector was an RCA connector, so this was a pretty non-standard piece of audio gear, but it served me well.

• • •

The Cadillac of projectors was the Elmo 1200 series. The Elmo ST 1200D, which was available for general release in 1976, came with a 150-watt bulb, and projected a bright, sharp image. Straight out of the box, the Elmo GS 1200, which hit the streets in 1977, came with a standard 200-watt halogen bulb, so it was immediately a great projector for larger rooms. You could even convert the GS 1200 to house a 300-watt, high-intensity, arc projection lamp. This conversion allowed you to project in a venue of almost any size. Sure, the conversion ran in the neighborhood of $1,000, and you needed an entirely separate power unit for the bulb alone,

but if you had the cash, it was a pretty bold maneuver. In 1980, Elmo developed the GS 1200 Xenon model that came preloaded with a 250-watt Xenon bulb. The 1200 series projectors were indeed fine machines and they had the price tag to prove it. The GS 1200 Xenon sported a whopping $3,700 price tag, while the standard GS 1200 would set you back $1,500. They lost no value on the resale market, and whenever I came across them used in the 1990s, they easily ran for over $1,000. That was a lot of money for a piece of small-gauge equipment, and I salivated like a hungry dog in front of many display cases housing this beast.

The other major benefit the Elmo projector had was its ability to play and record two tracks of sound. One of the peculiarities of Super 8 sound involved the magnetic sound stripes on the film. When Kodak first developed sound Super 8 film in 1973, they adhered a magnetic stripe to the film that allowed you to plug a mic into your camera and record sync sound dialogue right onto your film.

But what if you wanted to add music or effects? How would you do that?

Sound film stock actually came with two separate sound stripes. You recorded on the stripe opposite the film's perforations. The purpose of the second stripe, located alongside the perforations, was to simply balance the film out during projection. If only one side of the film had a stripe, the difference in thickness would cause projection and focus problems. Kodak solved this mechanical issue by adding that second stripe, also referred to as the "balance" stripe, because it balanced the film out during the projection process. Brilliant.

This was awesome. Or was it? Here's yet another uniquely Super 8 conundrum. If you had an Elmo 1200 and recorded sound on the balance stripe, you would, for all intents and purposes, be wholly dependent on that projector to screen your film. Again, most projectors couldn't play the balance stripe, so if you screened your film on another projector, your audience would only hear half of your soundtrack.

• • •

EQUIPMENT

Projection Problems

IT'S EASY TO TAKE for granted how good video projection has gotten in the last number of years. From quality video projection in classrooms to your parents dispensing of their conventional television set in favor of a stellar home-video projection setup, we are living in a high-lumen universe.

Trust me when I say this wasn't always the case. Getting good quality projection out of a film projector used to be an art in and of itself. Countless films and tv shows have parodied the frustrated 1960s school teacher helplessly trying to thread a 16mm projector in order to show her class the wonders of mitosis or strip mining. To be fair, threading a 16mm projector could be intimidating. However, most Super 8s, particularly sound projectors, had an automatic threading system that was fairly easy to use.

The big challenge of Super 8 projection was getting a bright enough image in a large space. When I first started out, this issue hadn't occurred to me. My guiding principle for projection stemmed from the realization that the farther you set the projector away from the screen, the larger the image became. In my mind, a big image was a good image. What didn't occur to me was that the farther you set the projector away from the screen, the dimmer the image became. This makes total sense, since the light source was moving farther away from the projection screen. This just never crossed my mind, and no one bothered to tell me. Perhaps it never crossed my mind because most of the time I was using my projector in the confines of small spaces. I projected in my living room and in my bedroom. If I was stepping out, I projected in friends' backyards to spice up a barbeque. I occasionally went out in public, screening films in dive bars before bands would play. In other words, small spaces, short throws, bright images.

When I first moved to San Francisco, I found a group of filmmakers whose work, like mine, crossed over between the experimental film world and the punk rock scene. There was a local space called Shred of Dignity. I had heard about it, saw some flyers, but never went to a show there until a friend asked if I wanted to show some films during a music bill. The flyer for the show was a black-and-white Xerox festooned with Geraldo Rivera's beaten face, his nose covered in bandages due to his recent on-air beatdown at the hands of some Nazi skins. Bands like Econochrist and Bazooka Joe were on the bill. It was a classic five-band, hardcore bill.

I got there and the space was huge. I mean HUUUUUUGE. A monstrous, industrial-sized warehouse. My eyes widened. The sound board was at least one hundred feet away from a giant cinder block wall that would be my screen. I parked my projectors next to the sound board, and we did a sound check while daylight was still streaming into the space. Sure the image was dim during sound check, but it would be dark by the time I screened my film. Once the films started screening, I was absolutely perplexed. You could barely see the image on the screen. The skinheads, the punks, and the girls with their Chelsea haircuts sat down to watch, but what they saw was anybody's guess. The image was practically nonexistent. The one hundred-watt bulb on my projector simply couldn't handle the throw. I kept thinking that there was something wrong with the projector, that the bulb was dying. I was so clueless. Honestly, it took me a couple years before I figured out what had happened, that I had placed the projector too far away from the screen. I was projecting a multiplex-sized image from a projector that could handle a small screening room at best. At least the films sounded good. That's something…right? ∎

Jet Set Projection

"WANT TO GET HIGH on Sophia Loren? Liz Taylor? Liv Ullman? George Segal? You can. Super 8 is riding with the big birds, flying to exotic places like Singapore, Kuwait, and Pakistan—adding *A Touch of Class* to jet travel, not to mention *Godspell*, *40 Carats*, and *The Way We Were*."

Inflight's Impak Super 8 cassette-loading projector for airline use.

We've become pretty spoiled when it comes to in-flight entertainment options on airlines these days. Back in the day, some airlines showed movies to their passengers. Inflight Motion Pictures and Trans Com were two of the companies chiefly responsible for providing the airlines with movie titles and projectors for their flights. Inflight had been providing 16mm films to the airlines since the early 1960s. In 1970, Trans Com developed a Super 8 cartridge-loaded projector that could hold 130 minutes worth of viewing material. Super 8 was making a play to take over the airline industry. By 1974, fifteen international airlines were using Super 8, pulling from a catalog of about "100 feature-length movies." The Trans Com projector featured a 300-watt Xenon bulb, while Inflight developed "a Xenon 500-watt arc lamp." Holy smokes! These companies weren't messing around. My favorite part of the pitch? According to Trans Com, "The weight savings alone on its Super 8 system for 747 jets would amount to between $42-$254,000 a year per airplane." ∎

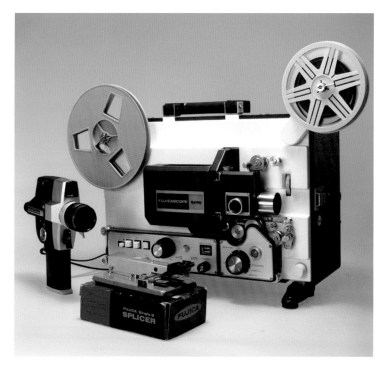

Fujicascope SH10 projector, 1970 with Fujica Single 8 P400 camera and Fujica Single 8 splicer.

But for all the projection horror stories and problems facing a Super 8 filmmaker, with the right projector, in the right screening conditions, the Super 8 image could look beautiful and wow an audience.

Using a middle-of-the-road projector like the Chinon in a microcinema could produce a glorious image. Watching your film on the Elmo in a microcinema was a sight to behold. Using the Elmo in a big theater, the Super 8 image filling a large screen was just as rewarding as watching a 16mm film.

There was also something wonderful about the portability of Super 8 projectors. Mine was no heavier than a typewriter. You could throw it under your arm and walk to a screening. You could toss it in your bike basket, head over to a friend's house and have a backyard BBQ screening. It reinforced the guerrilla aesthetic and the crackling immediacy so integral to the Super 8 universe. ∎

CHAPTER 6:

EDITING

EDITING

Arguably, editing is the biggest technical challenge of working with Super 8.

First off, Super 8mm is small. The size simply makes Super 8 hard to handle. Editing film is a hands-on, tactile endeavor and working with such tiny materials is difficult. Think about handling an eyeglass screw versus working with a more standard-sized screw. That is the difference between handling Super 8 and 16mm. Not only is the film smaller, but all the associated editing tools are equivalently undersized.

As already mentioned, working with reversal stocks has its challenges. The fear of scratching or damaging your film is ever present. The need for a clean workspace and clean equipment is paramount. You also need a steady hand and meticulous work habits to ensure you aren't damaging the film as you move through the editing process, from viewing to splicing to handling your outtakes.

Working with reversal also affects how you make edit decisions. Super 8 is spliced together with either edit tape or film cement. If you use tape and you make a cut you don't like, you can simply pull the tape off the splice and recut your sequence. Cement is more permanent and involves scraping the emulsion off one of the frames you're splicing. Therefore, if you make a cut you don't like using cement, you can't undo that edit without losing at least one of the frames at the edit point. With tape splicing you have a greater margin for error and experimentation.

Whether you use cement or tape, you tend to leave your shots long as you take a first pass through your film. It's always easier to trim frames out of a shot than to add frames back in. If you splice two shots together and don't like the rhythm and flow, it's easier to take the splice apart and trim away a bit from either of your edit points. If you cut your shots too short, it's much more difficult to add a frame or two back in. You'd first have to pull apart the original edit. As already mentioned, with cement you lose one of the frames at your edit point. If your goal is to extend the shot by a frame or two, this means you have a jump cut situation on your hands. If you are tape splicing, you won't have to lose any frames, but your film will visually suffer. Generally speaking, you can get your tape splices to appear fairly seamless. However, two splices in very close proximity will often be noticeable. Let's say you are adding just one frame to your shot, you will then have two physical cuts and two tape splices sitting next to each other in your film. You can do it, but there will be a good chance the audience will notice the cut and it will disrupt the invisible nature of the edit. To be frank, it's a pain in the ass.

Let's back up and talk about the tools and the process of editing Super 8.

To edit a Super 8 film you need a viewer (sometimes referred to as an editor), a splicer, and editing tape or cement.

When you start the editing process, you use a Super 8 viewer to look at your film. This allows you to move through your film slowly, analyzing it down to the frame level. Viewers are motorized or non-motorized. Motorized viewers allow you to look at your film at speed. Manual, hand-cranked viewers do not play at speed. With manual viewers, the speed at which you see your film is based on how fast you crank the handles. At best, you get an approximate sense of the rhythm of your film. With either the motorized or manual viewers, you can move your film one frame at a time, which allows you to make frame-accurate edit decisions.

Once you are ready to make a cut, you need a splicer to physically cut your film. Splicers have several essential parts. They contain a razor blade that cuts your film, and they have

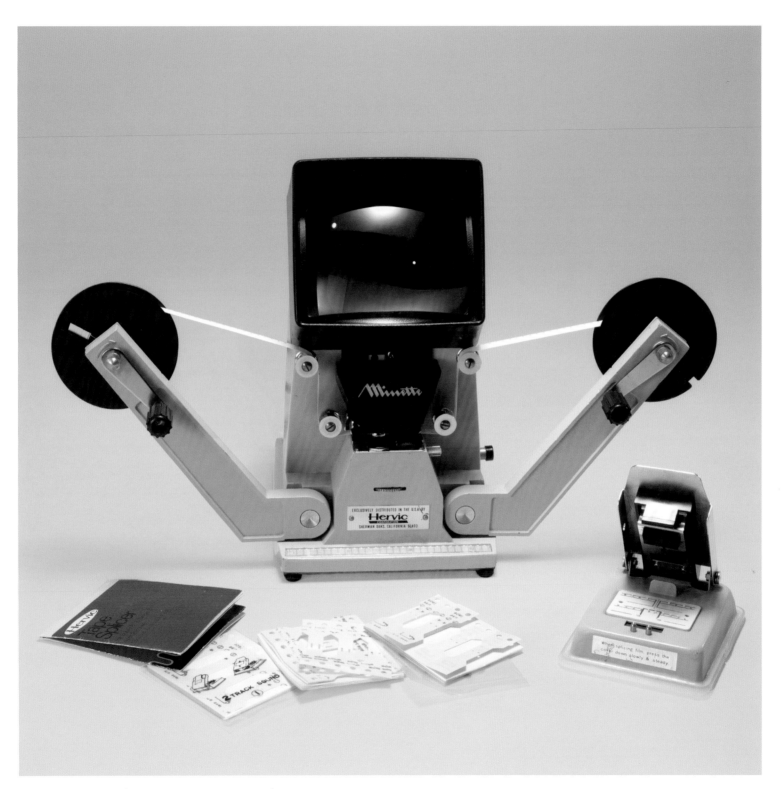

Minette S-5 Viewer, $85 in 1972. Hervic Splicer, $20.40 in 1979.

a platen with registration pins to hold the film in place while you make your cut. You place the film's perforations on the registration pins, which hold the film steady while you cut your film with the blade.

Cement and splicing tape, also referred to as edit tape, are the materials that hold your shots together.

Viewers

Similar to the camera situation, even during the medium's peak, Super 8 viewers varied wildly in terms of their quality. There were a ton of companies manufacturing them, many of these geared toward the home-movie market, which meant they were mediocre. You could readily find manual, hand-cranked viewers. Though screen size varied model to model, a 3 3/4" x 4 3/4" screen was a pretty standard size on which to view your image, which meant that you were looking at a pretty small image. Almost all of these standard viewers used 10-watt bulbs. These were automobile brake light bulbs, which made them the most inexpensive piece of film equipment available. But inexpensive doesn't mean good. With these systems, the image you were looking at was often very dim. The light from the bulb passed through a condenser lens, a prism, and a set of mirrors on the way to the viewing screen. The precision of this system was crucial in determining the sharpness and brightness of the image that you viewed. Some of these viewers displayed very nice images. On the other hand, with many of the cheaper viewers, you were often fighting and squinting to gauge the quality of your image. It should be fairly obvious that a small, dim image, playing at the wrong speed, makes editing a challenge.

Additionally, the majority of Super 8 viewers were silent. If you were shooting and editing sync sound film, you had limited options for cutting Super 8 sound. There were several mid-range models, both motorized and manual, that came with a shoe (a mount) into which you could attach a separate sound head. Companies such as GOKO, Elmo, Minette, and Bolex produced stand-alone sound heads that you could attach to your silent viewer. The attachment came with a line-out jack into which you could plug a set of headphones. But even if you had one of these sound heads, if you were hand cranking your film through a manual viewer, the sound would never play at speed. Regardless, this allowed you to cut for sound and was a workable solution for many filmmakers.

Several companies produced motorized sound viewers, and these were in high demand. The top-of-the-line viewers were made by GOKO. The GOKO RM-5005 and GOKO RM-8008 were amazing machines. Not only did they have sound functionality, but their optics were superb. Having one of these at your disposal transformed the postproduction process. The GOKOs were motorized and could play film at either 18 or 24 frames per second. They came with a strobe attachment, which allowed you to assess whether the device was moving at a true 18 or 24 (much like the strobes on high-end turntables at the time). If the motor was a little off, the GOKO had a +/- speed control setting so you could tweak the speed of your film transport if necessary.

Best of all, the image was sharp and bright. You could really see what was going on with your film. It should be said that even with the GOKO, editing with more than one person was a challenge. Though the GOKO had a screen of 4 1/4" x 5 1/2" and displayed a bright image, you had to be sitting directly in front of the viewer to see the image. If you were sitting off axis just a little bit, you could barely see anything. So, if you were working in a team of two or more, the edit process slowed down considerably.

The other beauty of the GOKO was the array of sound capabilities it offered. It played the main stripe. It played the balance stripe. You could record onto the main stripe. You could record onto the balance stripe. You could record sound-on-sound functionality on both stripes.

One of the big challenges of Super 8 filmmaking was post-production sound. You could record sync sound dialogue in production if you had a sound camera and sound stocks. If you had a good quality mic and if you managed to blanket the noise of the camera, production sound could be of very high qual-ity. Given the superiority of magnetic sound to optical sound (utilized for most 16mm prints), I would argue that in a perfect world Super 8 sound could exceed the quality of 16mm sound.

But postproduction sound was an Achilles' heel of the Super 8 production process.

If you wanted to add sound to your film in post, you needed a device that allowed you to record sound onto the stripes of your film. Many sound projectors had that capability, as did a handful of sound viewers.

For many, recording postproduction sound through a projector was the route taken, simply because projectors with these capabilities far outnumbered viewers with these capabilities. That said, if you had an option between adding sound via a projector or via a viewer, the viewer was an easier and more accurate route to go.

You could manipulate the viewer one frame at a time, which made it easier to achieve frame accuracy with your recording. Most projectors did not have that capability, so trying to hit a sound cue while playing your film through a projector was a cumbersome affair. You were ultimately hitting (or not hitting) your cues on the fly. With viewers you could nail it, or at least only be off by a frame or two.

Adding sound to the main or balance stripe was a relatively straightforward affair with the GOKO. The viewer had a mini input. You could line a tape recorder, a turntable, a mixer, a 4-track, a mic, or any other audio device directly into the viewer. You had level controls, VU meters, and you could monitor the sound with a built-in speaker or with headphones.

With the GOKO, you could enact simple operations, or complex operations. Let's say you had shot your film silently, but on sound stock. When you completed your edit, you could add a simple soundtrack on the main stripe of your film. All you had to do was load the film into the GOKO, cue up your sound on your turntable or cassette, start the GOKO, start the turntable, and press record on the GOKO.

You could get much more complex as well. Let's say the majority of your film was shot with sync sound, but there was a scene that was shot silently, and you wanted to add sound to just that scene. You would take that scene, isolate

GOKO RM-8008, 1978, $659

Zeiss Moviscop S8. Zeiss Ikon Movipress-Super 8, 1972, $79.

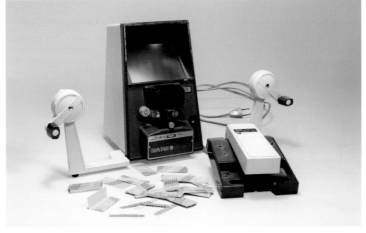
Suntar 303 Dual 8 Viewer. Kodak Presstape Universal Splicer, $15.75 in 1979.

CASE STUDY

My First Film: A Cautionary Tale

IN 1985, I WAS STUDYING at the University of Michigan. My first film was a pretentious, black-and-white art film about libraries. For a first effort, the film actually looks pretty good. I filmed in libraries, cemeteries, and elementary schools. I think I even filmed a bowl of oatmeal. When it came time to edit, I had no idea what I was doing. The school's editing equipment was pathetic, and to be quite frank, I don't remember getting any instruction on it either. I was facing a real self-taught moment.

I sat down at my desk to edit. The viewer worked well enough and I was armed with a Kodak Universal Splicer and presstapes. Unfortunately, the splicer's registration pins were worn down to a nub, bordering on nonexistent, and serving no utilitarian purpose. Without registration pins there was no way to hold the film in place while I made cuts or spliced my shots together. The bigger problem was that I didn't know they existed in the first place. Had I known, I might have requested another splicer from the film cage. Instead, I puzzled out what to do without these necessary guideposts. I carefully cut the film, hoping to make a straight cut at the frame line. I then held the cut film in midair, while simultaneously peeling the backing off of the presstape, then lining up the tape's sprocket holes with the holes on the film, and finally, I would affix the tape to the film in mid-air. I usually managed to line up the holes. If I succeeded, three frames of the presstape were attached to the first part of my cut. Another three frames of tape dangled in midair, waiting for the second shot of the edit to be attached. I then proceeded to cut shot number two, and in midair, affix the second piece of film to the dangling tape. If everything lined up, then hooray! If the sprocket holes on the tape and film didn't line up, I'd pull the whole thing apart and start over. It was a nightmare, but at least the film never fell apart. ∎

My Second Film: A Cautionary Tale

Note the tell-tale airbubbles of the Kodak Presstape rippling through the shot.

BY THE TIME I MADE my second film, I was clued into the fact that my first film was cut in a procedurally messed up fashion. Armed with this knowledge, I couldn't wait to start cutting my new film. But again, I was faced with a puzzle.

My first film was silent, and so the fact that the Kodak presstapes covered the entire width of the film was not an issue for me. My new film, however, was a soundie. I couldn't quite figure out what to do with these splice tapes that were going to cover my film's soundtrack. It made no sense to me. Nobody told me there were splices made explicitly for sound film, and it never occurred to me to ask. In my world, the Kodak presstapes and Universal Splicer were the only editing options available. Undaunted, I pressed ahead using the wrong tools for the job. My solution? Cut the presstapes length-wise with a pair of scissors, cutting away the portion of the tape that would cover the sound stripe. The dialogue was now free to come through loud and clear. However, an unsightly six frame vertical gash slices through my film at every edit point. The film in question—*Skate Witches*, my big internet hit. Be sure to watch the film with an eagle eye to see how that all plays out. ∎

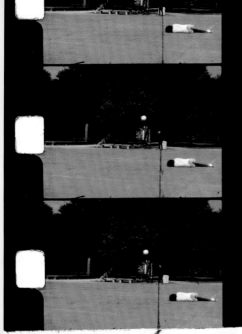

Can you spot how many things are wrong with the edit? 1) There's a big white flash at the edit point where the tape edit has stretched. 2) The blade used to cut the film was not properly aligned as evidenced by the line of the edit running at a slight angle. 3) The vertical line running through the frame because I cut the splice lengthwise to leave the soundtrack uncovered. 4) Note the fluctuating exposure at the head of the second shot. This was the result of poor control of the auto exposure function.

it on its own reel, and add white leader to the head and tail of the shot. This was an important step. You wanted to engage and disengage the GOKO's record button when the leader crossed the sound head. The act of engaging or disengaging often carried an unwanted mechanical buzz that you did not want on your film. Also, by isolating the scene, you ensured that you wouldn't accidentally record unwanted audio over another part of your film.

You'd cue up your sound on your playback device and set it in pause mode. You'd string the film up on the viewer with the white leader in the film path. You'd then start the GOKO playing, engaging the record button while the white leader was in the path and crossing the sound head. As the film played you could clearly see the demarcation point between your white leader and the first frame of your film. When the first frame of the film hit the sound head, you would disengage the pause button on your playback device so it would play, and voilà, you'd be recording sound. When you got to the white leader at the tail of your shot, you would disengage the record button and you'd be set.

You could do this on both the main stripe and the balance stripe. However, not many projectors would actually play the balance stripe, so it's unlikely you would actually use that functionality. The challenge became how to create a multi-track sound mix on a Super 8 film.

If you shot your film silently, it wasn't much of a problem. You could create a multi-track mix on a 4-track or 8-track audio mixer, and easily line that into your viewer or projector to record that multi-track mix onto the main stripe of your film.

But what if you had sync sound dialogue? How would you add music or effects underneath that sync sound? That's where the high-risk, high-reward sound-on-sound function came into play.

Sound-on-sound was a feature found on some projectors and viewers that allowed you to mix a second track of audio onto the main stripe. This would allow you to add music or effects underneath the sync sound you recorded on your main stripe. While this was a great feature in theory, it was a terrifying feature in practice.

Super 8 sound stripes were magnetic tape, just like cassette tapes. And just like cassette tapes, you could record on them and you could record over them. In other words, if you weren't careful, you could accidentally erase your sync sound dialogue. If for some reason you had your projector's record button engaged while projecting your film, you would unfortunately erase all your dialogue. You'd have to be pretty careless to do that, though I'm sure there are some horror stories out there.

Sound-on-sound functioned as a low-end, two-track mixer for your main stripe. Your viewer or projector had a sound-on-sound dial that you could set to different levels ranging from zero to ten. The zero to ten range represented the ratio of new sound to original sound in your two-track mix. If you set it at ten, you would record all new sound to your track. If you set it at zero, you would record no new sound to your track. If you were trying to add some music beneath your dialogue, you might set your sound-on-sound dial to two. The idea being that your new sound would occupy 20 percent of your track or would be 2/10 the level of your original sound. That might be a ratio of new sound to original sound that worked. Then again, it might not.

The frustrating challenge with sound-on-sound is that while you were recording, you never heard the new sound you were adding in relation to the original sound that was already there. In other words, you never heard the level of the music or effects compared to the level of the dialogue until after you had completed the recording. Worse yet, if you didn't like the mix, you couldn't erase the new sound you just added without erasing the entire track, including your original sync sound dialogue.

So, you had one chance to get it right. If you recorded your new sound too low, it would forever be too low, audiences always fighting to hear that music track. If you recorded your new sound too loud, it would forever make your sync sound dialogue hard to hear. You had one shot.

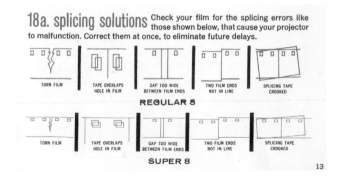

To compound your troubles, if you were trying to place an effect or audio cue on a specific frame, you had one shot to place it in the right spot. Remember, you were doing this all on the fly, so it wasn't easy.

You had to practice. If I knew I was going to be adding multiple tracks of audio to a given scene, I always made sure to shoot multiple takes of that scene. That way, I could do test mixes on outtakes. Quite often, however, I had only two outtakes. If on my first audio mix test, I set sound-on-sound to three and it was too loud, and on my second audio test I set my sound-on-sound to two and it was too low, then I would pick some point between two and three for my final mix and cross my fingers that I split the difference acceptably. It was a real high-wire act, but, if successful, you could get some good results.

As a result of these types of challenges, sound mixes on Super 8 films tended to be a little simpler than in other formats. For what it's worth, you could add a third layer of sound via sound-on-sound, but the audio quality would noticeably decrease, everything would become murkier, and ultimately it wasn't worth it. On the GOKO 5005 and 8008, you could even add sound-on-sound on the balance stripe, but as already mentioned, accessing the balance stripe in projection was never a given.

Regardless, this is how many people worked to create sound mixes. But Super 8 was a truly idiosyncratic medium, with no two people approaching a challenge in the same way. Other folks had different methods for dealing with their sound mixes. Some folks figured out how to pull their sync sound off the main stripe, bring that sound to an audio mixing system, create a complex sound mix, and then line this new recording back into a viewer or projector and rerecord their sound back on their film. The real challenge would be how you could guarantee that you could put it all back in sync. This seemed particularly insane to me, but creating high-level work in a low-end medium was all about ingenuity and derring-do. What might seem insane to me was particularly reasonable to someone else, and vice versa.

Splicers & Tape

There were a variety of of splicing systems on the market, and like all things Super 8 there was the good, the bad, and the ugly. Let's start ugly. The most ubiquitous splicer out there was the Kodak Universal Splicer. You could find these in most camera stores. This splicer utilized Kodak presstapes. You could readily find these tapes as well, and as luck would have it, they were relatively cheap. In the 1990s you could find them for $1.50 for a packet of twenty tapes. Not only were the presstapes cheap, but a splice made utilizing these tapes was fairly indestructible. They made a solid splice. That said, they were ugly splices. They covered six frames of film and were pockmarked with air bubbles. The goal of splicing is to hide your cut and make it as invisible as possible. There was no hiding a Kodak presstape splice. They destroyed any illusion of seamlessness. Additionally, these splices covered the entire width of your film. This made them useless for sound films because they would cover the sound stripe. If you were editing silently, there were several guillotine splicers on the market that could easily be found, and would be preferable to the presstape method. If you were cutting a sound film, you'd have to find another system.

By the time the mid-1990s rolled around and equipment was becoming harder to find, the next most readily available splicer out there was the Hervic Splicer. At the time, the Hervic cost more than the Kodak—$40 for the Hervic compared to $20 for the Kodak. The Hervic splice tapes were also costlier, with a packet of forty tapes going for more than six dollars. The Hervic splicing tapes came in two different configurations. One version was designated for "one-track sound," leaving the main stripe untouched, but covering the balance stripe. The other version was configured for "two-track sound," leaving both sound stripes open. Either way, they were an excellent choice for cutting sound films. They covered only two frames of film, which made them much less noticeable than the Kodak splices. You could easily tamp the tape down and eliminate air bubbles as well, but you still had to be careful. By virtue of covering two frames of film, Hervic splices were prone to stretching over time. If the splice stretched, a gap opened up between the pieces of film at the edit point. During projection this meant a disconcerting hop at the edit point, or a noticeable white flash, as the projector's bulb shone through the gap when the edit hit the film gate in the projector. I even had a Hervic splicer that started making cuts in the middle of the frame, not at the frame line, which was supposed to happen. Having cuts occur at the frame line was essential for masking the cut. Yikes. The indignities of the Super 8 filmmaker were endless.

The top of the line splicer was the Würker, which showcased German edit engineering at its finest. The Würker was harder to come by in America. By the 1990s, very few camera stores carried them, and quite often they were a special order, costing in the neighborhood of $100. The tapes were harder to come by as well, not to mention costing approximately $14 for a pack of fifty. The official Würker splices left both sound stripes uncovered and came in two or four frame configurations. I tended to edit with the four-frame splice. Though it covered more territory and was therefore more noticeable, I felt it a more solid splice that would hold up better over the long haul, and be less susceptible to stretching. A splice that stretched was not only a visual problem, but also caused havoc at the audio level. Let's say you laid down a piece of music over a series of cuts. If the edit points started to stretch, causing a miniscule gap at each edit point, then the sound would drop out at those points. Ghastly, but, sadly, all too common.

There were other companies making splicers and splice tapes, but Kodak, Hervic, and Würker were the most common. The Hama company made a splicer and tape that was similar to the Würker. It worked well, but had a cheaper feel than the Würker. It featured a lot more plastic and a lot less metal than the sturdy Würker. There were even a couple knock-off brands of splice tapes that you could use with the Würker splicer. This was helpful if you had trouble tracking down the official splice tapes. As always with Super 8, production practices were often dictated by what was available, as opposed to what you desired. ■

CHAPTER 7:

SOUND

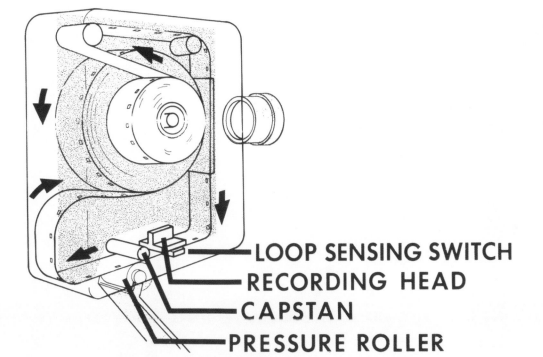

SOUND

From 1973 to 1996, Kodak manufactured sound film that came with two magnetic stripes capable of recording high-quality audio.

While shooting, you could capture sync sound live onto the main stripe or add audio onto both stripes at a later date using a projector or viewer that featured sound recording capabilities.

Recording sync sound was not that difficult once you knew what you were doing. That said, it was pretty easy to succumb to some pretty bad sound recording pitfalls. These mistakes often gave Super 8 an undeserved reputation as a format that delivered crap sound.

When you were recording single system sound, with the mic plugged directly into the mini jack on the camera, your immediate problem was the noise of the camera. Super 8 cameras were LOUD. The mechanical whir of the camera was heard on plenty of Super 8 films, and that aural intrusion immediately brought down the film's production values. As a result, the first thing you had to do was blanket the camera.

You could act like a professional and construct a "barney," which was a blanketing device. In their indispensible book *Super 8 in the Video Age*, Bob Brodsky and Toni Treadway give elaborate directions for building a barney for the camera by using "corduroy, vinyl, Velcro strips

and shag carpet." Any time you can get involved with shag carpet is a plus in my book! Canadian filmmaker Terry Pearce took it one step further, making a seal skin barney for his Nizo while filming Inuit life in the Northwest territories. To be fair, Pearce's barney wasn't blanketing sound, it was used to ensure the film inside the camera didn't freeze during shooting at -60 degrees Fahrenheit.

I tended to be a bit looser on my productions and would simply bury the camera under mounds of sweaters and blankets. This usually meant I was also covered with those blanketing devices while shooting. Another essential trick was making sure the microphone was as far away from the camera as possible. When recording sound, you always want the mic as close to your subjects as possible, but with Super 8 you also needed to ensure it was a good distance from your camera. The farther away, the less likely you would hear the camera noise. In other words, twenty-five feet of XLR cable was an essential tool of the shoot.

Interestingly, every box for a single system Super 8 camera prominently displayed the mic sitting atop the camera. While these photos look pretty awesome (see pages 19 and 36), this is exactly the type of setup you wanted to avoid. The Kodak Ektasound 240 series was by far one of the most ridiculous looking cameras ever designed. The mic was permanently attached to the camera, jutting suggestively out from the

Note the distance between the film gate and the sound head.

Bolex Striping Machine N8/S8, $319.50 in 1973.

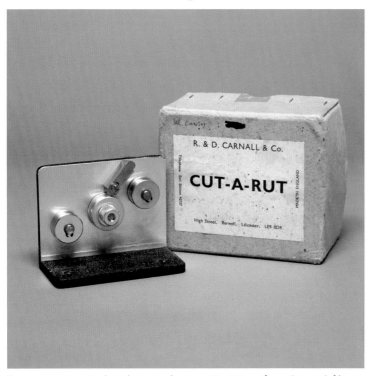

Removes unwanted or damaged magnetic stripes from Super 8 film. $20 in 1974.

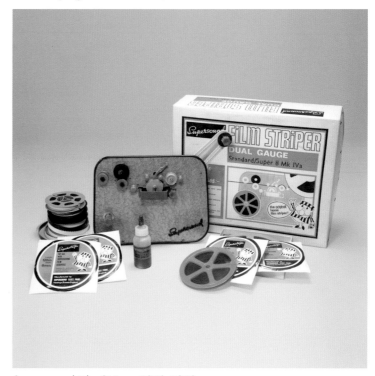

Supersound Film Striper, 1970–1979

camera's handle. The ad campaign even cloyingly asked you to "find the microphone in this picture…"

The next challenge was the layout of the actual film. In the camera, the Super 8 sound head was located 18 frames ahead of the film gate. Though you were recording in sync, the sound was physically located eighteen frames ahead of its corresponding image. The projectors were set up with the same layout. So, when you projected your film everything you heard and saw remained in sync. This layout, however, presented lots of headaches when it came to editing, which had to be accounted for when you were shooting. To avoid problems later, you needed to make sure the camera was rolling for eighteen frames (one second) prior to anyone in your scene uttering anything of importance. If your actors started talking right when you rolled the camera, those first utterances would physically live on the tail end of the preceding shot. This meant that when you were editing, you would have to cut out the first 18 frames of sound from that scene. *This sounds confusing because it is confusing.* If you shot properly, you'd be fine, but not everybody did, and as a result, you'd see lots of films where bits of sound had gone AWOL.

If you could wrap your head around these two concepts, you could record great sound. If you couldn't, you would have plenty of sound issues in your film. Because of the lack of postproduction options to sweeten the audio of your film, once the sound was messed up, it was difficult to fix. Films with bad sync sound were plentiful and were often held up as proof that Super 8's limitations made it an inferior medium.

If you shot on silent film, a necessity if you were shooting black and white, and wanted to add sound in post, you faced a whole other set of issues. Silent films could be post-striped for a mere twenty cents per foot in the 1990s. This was pretty cheap in the scheme of things. There were a number of labs and individuals who post-striped films. Oddly, this was not a service that Kodak offered. That's a shame, because in my experience, the quality of post-striped film was never as great as Kodak's pre-striped stocks.

When you post-striped you could get one or two stripes put on your film. There were several sound striping methods, but essentially a stripe was either laminated onto your film or applied in a liquid form that would leave iron oxide particles on the film once the liquid evaporated. Like every process with Super 8, the quality control was all over the map. Sometimes the striping was great, sometimes it was a mess. It was unclear if the quality of the mag stripe simply wasn't as good as the quality of the Kodak stripe, or if the problems arose from a lack of precision in the post-striping process. While it was possible to get good quality post-striping, a lot of the post-stripe output resulted in muddy, murky sound. ∎

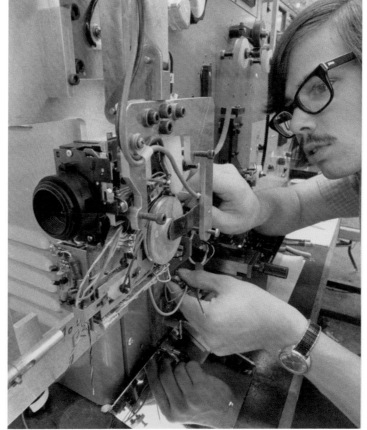

Testing and developing the Kodak Ektasound movie cameras, 1973.

CASE STUDY

PIPSQUEAK PfOLLIES: A Debacle of Print and Stripe

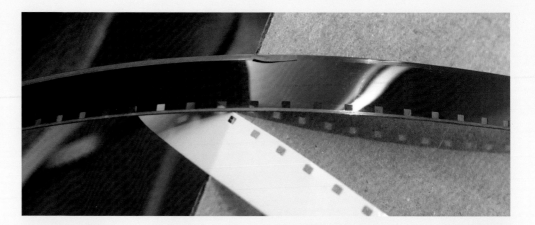

You can see the brown magnetic stripe peeling off the film. The main stripe is in dire shape, while the balance stripe is holding on.

IN FEBRUARY OF 1994, I completed a twenty-four-minute short called *PIPSQUEAK PfOLLIES*. It was a modern riff on a silent film. It featured a bunch of little kids terrorizing adults—a *Little Rascals*-gone-bad vibe. The film consisted of three parts. One section was shot with Kodachrome color sound film, and two sections were shot with Plus-X black-and-white silent film. The plan was to post-stripe the black-and-white film, lay down a score on that section of the camera original using the GOKO RM-8008, intercut the color and black and white, and then generate prints from there.

Laying the score down was an incredibly complex process. You'd never know it by looking at the film now. Today this would be a cakewalk, but given the Super 8 technology of the time, it was ambitious. I transferred the film off the GOKO by shooting the GOKO screen with a Hi8 camera. I dumped the Hi8 to VHS and gave that transfer to my composer who scored the film, based on the timing of the VHS transfer. She recorded the score on a 4-track Fostex analog recorder. Our plan was to lay the audio from the 4-track to the post-striped film via the GOKO. What made this challenging was that in the analog world there was no guarantee that the GOKO, the Hi8, the VHS, or the 4-track would consistently play back with the exact same timing. Even a minor fluctuation in timing would be significant over the course of a twenty-four-minute film. Bad timing would be noticeable, because the score contained several precise audio cues. Undaunted by the odds, we decided to give it a go.

The first technical challenge was striping the film. There was a local Northern California lab, W. A. Palmer that provided this service. They had always made my prints and had done a good job for the most part. Occasionally, I had some problems with sound quality on their prints, so I was amenable to exploring other sound striping options. At the time there was some guy in town who had a sound striping machine. The Super 8 world was filled with guys like this. He had come vaguely recommended, so I decided to give him a

"The sound stripe began peeling off the film in huge chunks. This was a disaster."

CASE STUDY

try. I liked the notion of going to the little guy, as opposed to dealing with a lab. That was certainly part of the ethos of the day. This would turn out to be a mistake.

Once the film was striped, we embarked on laying the score back to the film. Surprisingly, recording the score to the GOKO was fairly painless. To perfect the timing, we had to do some minor tweaking with the +/- speed control on the GOKO, but we hit our marks and felt pretty damn proud of ourselves.

We played the film back a couple times in a projector to check our work, and very quickly, problems arose. The sound stripe began peeling off the film in huge chunks. This was a disaster.

No one could tell me the efficacy of restriping a previously striped film. I wasn't sure it could be done. First off we'd have to peel off the remaining stripe. Secondly, if there was any residue of glue still on the film, that would have to be cleaned off as well. Somehow, I felt we would just end up doing more damage to the film. I was reluctant to take this approach because we were working with the camera original, and as of yet, we didn't have a print or a video transfer which we could screen or use as a safety copy. At this point the film was unviewable.

Plan B was to get a print of the film, stripe that, and try to lay the soundtrack down again. Because the print stocks at the time were all color, any print of a black and white film would have a slight tint, but I was willing to accept that. PIPSQUEAK PfOLLIES was a nod to the silent film era and I could justify a tinted print as an old timey artifact. When I got the print back from the lab, I was appalled. I was told the tint would be blue, but what I got was a Pepto Bismol pink. The lab profusely apologized. Clearly they had used the wrong filter pack during the printing process. They owned up to their mistake and were glad to redo the print for me at no charge. I was happy to wait a couple more weeks for a new print.

But then I got a bizarre call from the lab. Their guy who did the printing had taken ill and was hospitalized. The lab didn't want to run any prints without his supervision and they wondered if I'd be willing to wait a couple more weeks until he returned to work. Tragically, he passed away, and the lab was left without someone who fully understood the printing process. They tried another print for me. The new print was blue, as originally promised, but it was way too blue. They agreed that the print didn't look right, but weren't sure what they could do about it in the short term, given the death of their long-time, beloved staffer.

I attempted a third print at a different lab. This print came back green and the quality of the sound stripe was terrible. We laid the score back to this track, and it sounded like the score had been recorded underwater. This was the print I used for my world premiere. It kind of stunk.

With the original unscreenable and the prints a disaster, I gave up on trying to get an acceptable Super 8 print of the film.

Ultimately, I got a 1", high-end video transfer of the film and blew that up to 16mm. Sure it cost thousands, but that was sometimes the price of working in Super 8. At least it sounded good! ∎

CHAPTER 8:

PROCESSING

In the San Francisco Area: There is now a

KODAK CONSUMER CENTER IN PALO ALTO.

Along with your photo dealer, we're here to help you. If you have any questions about warranty service or about repairs on Kodak cameras and projectors, we're now at 925 Page Mill Road, Palo Alto (phone 493-7200), as well as in San Francisco at 3250 Van Ness Avenue (phone 776-6055), and in San Ramon at 9100 Alcosta Boulevard (phone 828-7000). The hours are 9 to 5, Monday through Friday.

EASTMAN KODAK COMPANY

PROCESSING

Filmmaking is a photochemical process.

Light sensitive silver halide crystals are embedded in the film's gelatinous emulsion. When you shoot, those crystals get exposed to light. After shooting you send your film to a lab where the film is dunked in a chemical bath. The chemicals interact with the silver halide crystals on your film, bringing the image to life. This process makes you reliant on film labs.

I can't overstate how integral labs are to the filmmaking process. The freshness of their chemicals and how they treat your film are paramount in guaranteeing your film's image quality is as good as it can be.

In the 1980s, there were a tremendous number of film labs processing a variety of film stocks and film formats. As the 1990s wore on and gave way to the new millennium, labs started disappearing. As digital video took over the consumer and prosumer markets, the number of labs processing small gauge film took a big hit. By the 2010s only a handful of labs remained standing.

In the 1980s, most major cities had local labs that you could use for processing. A mid-sized city like Detroit actually had three film labs where you could process your film. In many cases these labs offered same-day service. You'd drop your film off by 9:00 a.m. and you could pick it up at 4:00 p.m. You could also drop your film off at places like Kmart, Sears, and Fotomat, as well as most photo shops and camera stores. They would courier your film to a nearby lab. Color or black and white, sound or silent, they would ship it off and you'd pick it up at the store later that week. In the 1980s it might cost as little as $3 or $4 to process a roll of film. In the late 1990s you could even drop your film off at Safeway or Costco!

Kodachrome was the one stock that was a bit of a tricky beast. Due to its complexity, many local film labs didn't touch Kodachrome processing. Until 1957, Kodak was the sole processor of Kodachrome, but after a court order, they were forced to share their processing techniques with non-Kodak outfits. "Kodak actually had to go out and essentially fund laboratories to be their processing competitors," says Kodak engineer, Roland Zavada. "Then we had to support those processing facilities with technical expertise until the industry gained a foothold and momentum of its own." Kodak had a constellation of regional labs spread around the United States. Rochester, New York serviced the East Coast; Dallas, Texas, serviced the South; Findlay, Ohio serviced the Midwest; Palo Alto, CA handled the Western States. You could ship directly to Kodak, but back in the analog heyday, you could readily find a local camera store that would ship to Kodak for you. Most local camera stores were feeding Kodak with 35mm stills and slides for processing, and it was easy enough for them to include motion picture films in their daily shipments. Also, you could buy prepaid processing mailers when you purchased your film. If you did this, after you shot your film, you'd pop your cartridge in the mailer, throw the mailer in a mailbox, and the US Postal Service would deliver your film to Kodak, who would process it and return ship it back to your house.

In December of 1997, Kodak ceased processing Kodachrome at its Dallas facility, which had been the last Kodak lab in the US to handle the stock. From that point forward, if you wanted Kodak processing, you were reliant on the Kodak lab in Lausanne, Switzerland, which carried the Kodachrome torch until 2006.

Once Kodak ceased processing Kodachrome in the States, if you took your film to some of the chain grocery and drugstores, they would ship your film to Fuji in Arizona, which was also processing Kodachrome. It's worth noting that the Fuji processing of Kodachrome looked horrible. The normally vibrant blues and reds were sucked of their soul, with Fuji adding an odd yellow sheen to the proceedings. I mention this as a reminder that your lab mattered. The chemicals they used affected the look of your film. It was important to find a lab you trusted and a lab that delivered consistently good work. Today, processing of the remaining Super 8 film stocks is handled by independent labs.

Now that we live in a digital world, filmmaker/lab relationships are mostly a thing of the past. But as anyone who cut their teeth on film knows, that relationship was a special one. It was not always pleasant, but it was a tangible part of the process. You built up friendships with the folks at the camera store who took your film. If you were dealing directly with a lab, you'd have conversations about when the chemicals were the freshest. If you saw any image irregularities, you'd make a phone call and have a conversation. Some of those conversations could get heated, but everyone involved in the process was working for the greater good, trying to get the most out of the celluloid image.

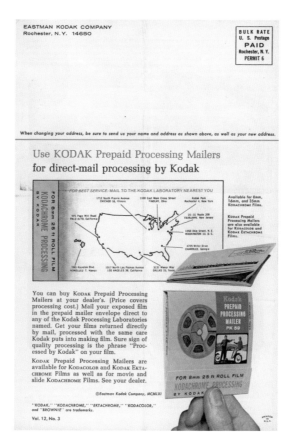

Just recently, a batch of film my students shot came back with some irregular flashing throughout some of their rolls. We had been using some older cameras, and it was unclear to me if the issues were the result of a light leak in the camera, or an issue at the lab. My students actually liked the resultant look, so they weren't particularly concerned. However, from my perspective, it was important to know if we had camera issues, and it was important to the lab to know if they had an issue with one of their machines. After going back and forth with the lab, sending them a transfer of the footage in question, they were able to trace the issue to one piece of their equipment that had a microscopic light leak. The lab was glad to identify this, so as not to send out any more film with such problems. At the end of the day, even though there was a problem, this helped me cement a personal relationship with the lab. In many respects, this is not an unfamiliar story of the filmmaking process.

The other exciting part of processing, which has no resonance today, is the moment of unveiling—the moment you get your film back from the lab and see it for the first time. What does it look like? Did it come out? Unlike digital video where you can easily monitor your image as you shoot, the Super 8 production process is quite different. What you see in the Super 8 viewfinder does not bear resemblance to the final product. Unless it's dark outside, everything usually looks pretty good through the viewfinder of a Super 8 camera. You don't see underexposure or overexposure. The only way you know if you are properly exposing, underexposing, or overexposing is by understanding how to read and interpret the f-stop values presented by the camera's internal light meter readings. Also, let's not forget that your color space, your grain structure, and your contrast ratios all differ stock to stock, yet through the viewfinder you don't see how any of these visual components play out. The notion that what you see is not what you get is a tough concept to wrap your head around, and has resulted in some poorly shot films over the years. At the end of the day, this means that the first viewing of your film, upon return from the lab, is really the first time you see your film. Even as a seasoned pro, confident in your ability to read a light meter, there is always an undeniable thrill in seeing your film for the first time.

Usually, I can't even wait to bust out my projector. I've unspooled freshly processed film in my office, in my living room, in the lobby of a camera store. Unspool it and hold it up to the light. Squint at the 8mm images. Is there image? Yes! Hooray. No! Boo. Technically this is bad form. Unspooling it by hand and hoisting it aloft toward a light source usually results in several feet of film dragging on the floor. The more you unspool and the deeper into the film you insist on going, the more film ends up comingling with dirt, dust, and other unsavory bits of flotsam that live beneath our feet on a daily basis. Caution be damned, it had to be done. Filmmaker Dave Markey recalls, "I remember counting the days down until I would get that little 50-foot reel back. Then I would anxiously get to that Fotomat kiosk on my skateboard and before I'm even home, I'm already holding the film up to the sun looking at it, looking at the image as I'm on my skateboard going down the street. I have fond memories of that and it's all very precious for me."

Only after this ritual would I bust out the projector. Hopefully it was night, the dark room providing the best conditions to assess the image. I'm sure all filmmakers have their habits, but I'd always watch the film by myself before inviting cast or crew over. I needed to know it was alright. If I messed up, I needed to gather myself to figure out my next move and to line up my excuses! But honestly, watching every new roll of film was exciting. That's a thrill that the digital revolution can never match. ∎

TECHNIQUES

Hand Processing

NOT ALL FILMMAKERS are dependent on the lab to process their film. There is a rich, DIY tradition of hand processing your own 16mm and Super 8. If you have the requisite chemicals, tanks, buckets, and a darkroom, you can eliminate the film lab from the equation, and exercise greater control over the entire filmmaking process. In the heyday of analog still photography, you could easily find a darkroom where you could take care of business. Given the prevalence of home, community, college, and high school darkrooms, this was a path within reach of many filmmakers.

When hand processing film, artists can choose to go down two distinct paths. You can aim for lab-like precision, or you can take a more experimental approach to the endeavor. If replicating lab procedures is the goal, you need to to get a tank with a reel to accommodate your film. In the dark, you remove your Super 8 from its light tight cartridge, thread your film onto the reel, and place it into the tank, at which point you are ready to add the chemicals that will turn your latent images into images ready for projection. During the prime years of Super 8 reversal film stocks you could hand process all the black-and-white films, as well as all of the iterations of Ektachrome. Kodachrome was the one stock that needed professional lab processing, though today people are shooting expired rolls of Kodachrome and processing it as black-and-white film.

But often, when artists hand process film, they are hoping to create a more poetic, atmospheric, or experimental look. The "spaghetti method" is a popular way to hand process film. Instead of reeling the films into conventional processing tanks, a filmmaker can literally hammer open the

EQUIPMENT

Hand processing 16mm film "spaghetti-style."

50-foot cartridge and unspool the film into a bucket, which is then filled with chemicals. When using the spaghetti-style, you can process multiple rolls of film simultaneously, which results in a whole host of visual artifacts springing to life on the film. In the bucket, strands of the film overlap, stick to each other, and as a consequence not every frame of film gets exposed to an even wash of chemicals. The quality of images and colors can fluctuate wildly and unexpectedly throughout the reel. Even a single frame of film might contain several different looks. This variance and these "imperfect" outcomes can yield beautiful and evocative images.

In 2010, the collective unconscious was at work, and a new movement of eco-processing took root. Film processing chemicals are bad for the environment, they cost money, and aren't always available when you need them. Filmmakers began exploring alternative ways to process film. Could you develop film with everyday materials that you could find at a hardware store or the local grocery? The answer is yes! Over the past ten years filmmakers from around the world have shared their recipes for hand processing. You can process your films with coffee or a local microbrew or freshly cut flowers straight out of your backyard. Dagie Brundert from Germany and Lisa Marr and Paolo Davanzo of the Echo Park Film Center have been at the forefront of these explorations. The results can be stunning. Hand processing is already a DIY endeavor, but creating your own processing brews ratchets up the do-it-yourself quotient exponentially. ∎

Polavision

THE POLAVISION CAMERA and viewing system was a disastrous attempt by Polaroid to create a movie version of their vaunted Polaroid camera line.

Polavision promised instant movies. Polavision used film that was "identical in width and perforation to Super 8" loaded into a cassette that was more akin to an audio tape than a traditional Super 8 cartridge. Polaroid also referred to the stock as "Polaroid Phototape." After the tape was shot, it was inserted into the Polavision player, a twenty-five-pound viewer the size of a portable TV set of the era. When the tape was inserted into the player, a "layer of developer was laid down on the tape as it rewound." Less than a minute later, after the "processing reaction" was complete, the tape was ready to viewed in the player. This sounds cool, but by all accounts the system was a disaster. The image produced by the camera was shabby and like most viewers, only a couple people could satisfactorily view the playback at any one time. RIP Polavision 1977–1979. ∎

The Supermatic 8 Processor

IN 1974, KODAK UNLEASHED the Kodak Supermatic 8 Processor, a do-it-yourself processing unit. "The size of a large Xerox machine [that] can be installed anywhere faster than a home washing machine," this automated system could process film in just under fourteen minutes. The target market was TV news and companies making industrial films. The $12,500 price tag made this beast a bit cost-prohibitive for the hobbyist. An ad for the unit makes the argument that "about ten Super 8 cartridges per day amortize the capital outlay, over a few year period, when compared with commercial processing costs." ∎

INTERVIEW

Frank Bruinsma: Film Labs in Modern Times

Frank Bruinsma drying Super 8 film.

FRANK BRUINSMA STARTED the Super8 Reversal Lab in 2000. Located in Den Haag in the Netherlands, the lab formed out of the ashes of the beloved foundation, Studio Één (Studio One). Bruinsma has spent years crafting his hand processing technique, achieving lab quality results with less wear and tear on Super 8 footage than conventional machine processing. As one of Europe's top labs processing Super 8, he has a great vantage into the world of European Super 8 filmmaking.

By 2000, the year you started Super8 Reversal Lab, sound stocks are gone and there's a lot less support for Super 8. What made you decide to launch a Super 8 business?
I had a big chance of going completely independent. Studio One was bankrupt, so they had to close the foundation. I had been a volunteer for [them] since 1993 so I knew all the equipment. I knew a lot of the filmmakers. I already processed a lot of Super 8 for them at Studio Één, but I also had my own lab at home. When I got the opportunity to buy half the equipment with processing tanks and mixing equipment, and of course with a bunch of contacts and potentially new customers, I said to myself, "If one wants to be independent, now is the time to do it." I never thought about if it was going to be successful. I never thought if it was the right time. I always hear the same things: "Super 8 is dying" or "Super 8 is coming back." People come into my lab and say, "Oh, it's a revival of Super 8." And I tell them, "Were you here eighteen years ago, you would have said the same [thing]." I don't think there is any good moment to start a lab or to stop. You just have to be lucky with what you're being offered. So I jumped into it.

Was Studio One solely focused on Super 8?
They did Super 8, 16mm, and they did film processing, film printing. They also had a striping machine. They rented cameras, they rented loop boxes, projectors. You could do-it-yourself process. I teach many young filmmakers, artists, and students from art schools how to process your own film and how to mix black-and-white chemistry. There were several opportunities being offered there.

And then you decided to make your lab exclusively Super 8. Why didn't you stay on with the 16mm?
I don't like 16mm. It's too beautiful.

What are the qualities of Super 8 that excited you?
Although I have quite a lot of experience in Super 8 shooting, it still surprises me, because Super 8 has many different emulsions that you can buy, especially those expired cartridges. That's not so available in 16mm. If you want some old Ektachrome 16mm, it's quite difficult to get your hands on. With expired Kodachrome or Ektachrome or Agfa Moviechrome in Super 8, I find it at almost every flea market I go to. I always thought, since I started filmmaking, that Super 8 had much more variety and was much more flexible. The 16mm was much more professional. Strange enough, I never considered 16mm to be an experimental format. I always thought it was just too too sharp, too detailed, and too good.

The Netherlands is a small country, so I presume you're getting orders from all over Europe?
That's correct. If it hadn't been for Europe and for the euro, I would not have survived.

Are there many other Super 8 labs in Europe?
There are a few of them. The biggest and best one is Andec in Berlin. They process 200 cartridges a day. Another one is called Cinédia, which is in Paris. They also process Super 8. And Gauge Film, which is based in Ireland, and my lab. Those are the four most important processing labs in Europe.

When you started the lab, what services were you offering?
I offered everything. I was going crazy doing every emulsion. Black and white, black-and-white negative, color negative, color reversal, cross processing. I hired cameras; I hired loop boxes, projectors. I teach people, so they could come over for an afternoon for four hours and I'd charge them ten euros and I explained to them how they could make Super 8, or how the camera worked. I did almost everything in the beginning because I didn't have enough processing work.

Were you doing prints as well?
Yeah. I bought the equipment from Studio One, and I still own it. I have a Takita Blow-Up Super 8 to 16mm machine, and I have a Takita Super 8 to Super 8 contact printer. I made a lot of prints for many artists in France, because you have Light Cone. Many artists from Light Cone needed prints for the screenings. I did many of those prints in those days. But you couldn't make money out of it, because it was a very long process. You first have to see what it looks like, then you have to decide to make a test strip, not only for the exposure, but also for color correction. Then you had to decide which was the best—if it should be a little bit too blue, or a little bit too brown, or a little bit too red. Perfect prints on Super 8 were very difficult to make. After making three or four tests, I made a final print. If I made a mistake in processing, I had to start all over again. I didn't make that many blowups because not many people wanted or needed blowups, but I did make thousands and thousands of feet of color prints.

In 2000 how many Super 8 print stocks were there?
One: 7399. I only used 7399 from Kodak, which is a color reversal print stock Ektachrome. I was very upset with Kodak that they stopped making that in 2004.

What are the services you're doing most now?
A year ago I stopped offering scanning services through Amsterdam because they raised the prices so high. I couldn't tell someone, "I can process your six Super 8 cartridges and with the high definition telecine I will charge you 400 euros." That was completely stupid. So I told everyone, "Okay I'm only processing film and if you want to get a scan, just ask someone else in the Netherlands because there are people doing it for two euros a minute." It was a big decision, but now I'm very happy. Since that period I shout-out to everyone that wants to hear it that I only focus on processing film. I never thought that would be enough to stay alive. Strange enough, I get more processing work than I ever did. I get films from all over Europe. I get old stuff, new stuff. You name it, I get it.

How many rolls a week are coming through your lab?
If you have current films stocks like Kodak Tri-X, I can do about a hundred a week. If it's expired film, then I might only do fifty a week. It all depends on the emulsion type. If I get 16mm for example, I have a limited amount of drying space because I dry my films on a rack like how you dry your clothes. I only have space for six 16mm films, so I have to wait until that's dry, until the next day, and then I can do the other 16mm films. That's my limitation and practice. I could process more 16mm film a day but I cannot dry it. I want to keep the turnaround time as short as possible. But sometimes I have only two cartridges of color negative, and I cannot make the chemistry with just two cartridges. I have to explain to people that I work on batches and I need at least eighteen color negative and twenty-four black-and-white reversal or thirty-six expired films before I start doing that.

Talk about your processing method.
I hand process everything and that's the basis of my success. With hand processing you don't need that much chemistry, but I hand process with all the tanks in a large bath which is temperature controlled.

"I get more processing work than I ever did. I get films from all over Europe. I get old stuff, new stuff. You name it, I get it."

INTERVIEW

INTERVIEW

For someone who doesn't really understand processing, how does what you do differ from what one of the bigger labs does?
It's more physical work. Basically a large, professional motion picture lab has to mix the chemistry. They need very large containers for that. I only need a small, four-liter glass bulb to make it. So, that's practical and simple. And the handling of the processing is that I take the film out of the cartridge just minutes before it's being processed. Major labs have to staple everything together on very large rolls, which is a higher risk of damage and scratches. That's my advantage. The process itself, in the machine, works more aggressive than hand processing, because a machine is limited to a certain amount of times and temperatures and running speeds. With my hand processing, I can change everything at every specific moment in the system during the day without that being trouble making for the first or the last films that I process. In general, if you would compare my lab to to a big lab the whole process is actually the same. There's no difference.

So the quality is going to be, relatively speaking, the same, and there is a chance that yours might be better because you have more control at various points in the process?
That's correct. That's why I'm well known. For more than a year, I process all E6 color reversal films for Andec. Now German filmmakers write me and they say that my film process is brilliant and they want to stay with my lab. Andec always used a film processing machine, so that's my evidence that hand processing can be better than machine processing.

Does anyone ever request spaghetti-style hand processing?
Never. No. It's the opposite. People are disappointed if they hear that I don't have a machine. That's why I don't write on my website that I use a machine or don't use a machine. Strangely speaking, human beings want you to use a machine. They only have confidence in the machine, right? Well, the opposite is actually more true because what I do has been done for more than a hundred years, and it's still

Super 8 Vitrine.

valid and it still gets very good results. I make test strips and those test strips are being made by Haghefilm and then being measured. My color negative processing is, technically speaking, exactly the same as their machine processing. Those people working at the lab in Amsterdam were very surprised that I can maintain that quality hand processing, because they say hand processing color negative must be much more difficult because the temperature is so critical. You would say it's impossible to get equal results, but still I get it.

Let's talk about some of the non-Kodak stocks available today. Can you talk about Wittner?
Mr. Wittner is one of the few people in Europe that rebrands film stock into Super 8 cartridges. At least he tells the buyers what's inside. So that makes it a whole lot easier. The cartridges that he produces in black and white are all very fine technically speaking. And the emulsions in black and white are very nice. You have an Agfa CHS II, which is a very fine black-and-white negative emulsion. If you would use that and then scan it in high definition you would get very beautiful black and white.

Is he cutting them down from 35?
Yes, he is. He has all the machines. But he's cooperating. In Europe we have Andec, ADOX, Orwo, and we have Wittner. They all participate in one way or the other with each other's products, because not all of them have all the equipment. They all need each other. Everybody needs someone to either cut, slit, or perforate, or divide rolls into small pieces, so they are all depending on each other.

Is there another company producing Super 8 film stock in Europe?
We have ADOX, which is recutting and perforating Agfa film. He also cuts and perforates another black-and-white stock, but he doesn't tell anybody what

it is. Everybody expects that it's very old black-and-white Agfa film. It's a drama to process. Just to give you an example: if you would process a Kodak Tri-X 7266 at five minutes at twenty degrees, you would need to process that ADOX film at seventeen minutes at twenty degrees. It's this very strange black-and-white film. And then we have Kahl Film, who restocks old Ferrania or old Orwo materials. He's very difficult to do business with because he stopped selling to consumers. He only wants to sell to businesses.

What type of projects are coming through the lab?
Most of it is family stuff. The users of my lab, that keep me alive, are the amateur filmmakers that do it because they love it, and because they do it already for thirty years and don't want to stop making Super 8. Then we have the artists. I work for a few artists that regularly shoot like ten cartridges a month. After the artists we have a small amount of people that do it professionally. Then we have the students, the smallest group of people, because they simply cannot afford Super 8 anymore. And they all want digital, so you have to add the cost for scanning film.

Are there still schools in Europe teaching Super 8 as a starter medium or has that gone away?
Everybody stopped it. All the arts schools, all the high schools. Nobody teaches film. Nobody teaches Super 8. Nobody teaches analog film photography. What is a little bit coming back is that a few of the art schools have set up darkrooms again for photography because there were so many requests. In the nineties, I taught in several schools on behalf of Studio One, but that is all gone. In Europe everything is digital, digital, digital. Everything is Black Magic and Apple. It's dramatic.

What does the future of Super 8 look like to you?
With the return of Ektachrome there will be a short revival in the sense that it gets more public attention and it might even get to the nine o'clock news, which is fantastic. I also believe that their bringing back Ektachrome helps keep labs alive because most people that shoot Super 8 want reversal stock. So that's very important. My fear is that Kodak might be disappointed in about five years' time that they don't sell as much Ektachrome as they wanted to. But then someone said to me, "You're looking at it the wrong way, Frank, because Super 8 is just a by-product. The main product of Ektachrome is still film." I never looked at it that way. He could be right, and I hope he's right. ■

CHAPTER 9:
PRINTS

| BATMAN | 8 HOME MOVIE | BT-2 | EPISODE 2 |

CASTLE FILMS — SON of FRANKENSTEIN — 8MM COMPLETE EDITION
No. 1033

CASTLE FILMS — THIS IS WAR? — SUPER 8mm COMPLETE EDITION
No. 866 B/W

FILMS — MEET Dr. JEKYLL AND Mr. HYDE — BLACK & WHITE For Use Only On SUPER 8 Projectors / BLACK & WHITE SUPER 8 COMPLETE
852

8mm © COLUMBIA PICTURES HOME MOVIE
ensed for non-theatrical home use only. All other use prohibited. Printed in U.S.A.

CASTLE FILMS — MEET THE MUMMY — 8MM COMPLETE EDITION
No. 861 BW

m SILENT   CLOCK CLEANERS   SUPER
W   1402

CASTLE FILMS — THE WOLFMAN — 8MM COMPLETE EDITION
NO. 1060

PL-1 PLANET of the APES   SUPER 8

PRINTS

One of the most frustrating experiences of the Super 8 process was getting prints of your films.

This difficulty opened up a can of worms when it came to exhibition. Should one show the camera original? Should one go down the rocky path of generating prints? Could one show a video transfer of the film?

Inevitably you transferred your film to video. By the early 1990s, you regularly submitted a video copy of your film to festivals as part of the application process. However, until the mid- to late 1990s, video projection at the microcinema and cinematheque level was often substandard. So even if you had a high-quality video transfer, you couldn't depend on video being a viable method of exhibition.

Theoretically, the best option available would be to get a film print. This would seem to be a no-brainer and was standard operating procedure for 16mm negative films. Unfortunately, for Super 8 films shot on reversal stocks, the quality of prints varied greatly. The majority of Super 8 prints were made by a method called "contact printing." Your original film would come into physical contact with the print stock in the gate of the printer. At that point, a beam of light would pass through the original film to expose that image onto the print stock. Even under the best conditions, prints made in this manner suffered generational loss. Your film's sharpness would decrease ever so slightly, while the grain and the contrast would increase ever so slightly. Also, colors from the original never fully matched the colors on the print. So even the nicest print was slightly disappointing and begrudgingly accepted.

By the early 1990s, the standard color print stock available for reversal films was 7399 (stocks were often identified by number), which presented challenges when it came to getting a true color match between print and original. The 7399 was an Ektachrome stock. To achieve optimal colors in the printing process, labs would use one filter pack when generating prints from Ektachrome originals, and a different filter pack when generating prints from a Kodachrome original. If you mixed film stocks within your original film, you were really screwed. The lab couldn't switch out filter packs mid-print, so you'd have them choose one filter pack and adjust the colors based on which stock you used more of in your film.

Black-and-white prints were even more problematic. There was a black-and-white Super 8 print stock called 7361, which was retired in the early 2000s. After that point, black-and-white prints were made on the 7399 color stock. Keith Anderson, co-owner of Yale Film and Video, told me, "When we have a color [print] stock, we can't completely filter out all the color. So, when someone had black and white, we would have a special [filter] pack to reduce as much of the color as possible, but still it would have a slight blue tint to it." In other words, if you were shooting black-and-white, you'd never get a true black and white print. "What made it even more complicated was when someone would throw in some shots of black and white and put in some Kodachrome and some Ektachrome." In other words, this was a situation featuring three different original stocks, all requiring different filter packs, but with the lab in the position of having to choose only one filter pack for the whole project. "We would tell people, 'All right, you've got a mixture of color and black and white. We have to [use] color stock and the black and white is going to have a blue tint to it. We're going to do the best we can."

There was a second type of printing process called optical printing. This process involved a projector and a camera

pointing at each other, with a small screen placed in between the two machines. The lab would project your camera original onto the screen, and it would be rephotographed, frame by frame, by the camera, which was loaded with the print stock. Some labs took this route, but according to Anderson, the process of making Super 8 optical prints was even more challenging, particularly when it came to maintaining focus on the print. Super 8's small image size complicated this practice. Also, if you were mixing different stocks, they each had different focal lengths, which was another knotty problem for the lab to contend with. With contact prints, you didn't have to concern yourself with focus, since, by virtue of the films touching each other, the focus between print and original was guaranteed. It should be said that some people blew up their films from Super 8 to 16mm, and if they took this path, the optical printer was used for the blowup.

One of the beauties of the 16mm negative print process was the creation of timed prints. When the lab made prints, they could go through your film scene by scene and make exposure adjustments. If you shot one scene underexposed, they would pump more light into the negative, so your print would come out looking properly exposed. Not only did they make exposure adjustments, but they could adjust red, green, and blue color values as well.

When making Super 8 contact prints, the labs didn't give you timed prints. They made what was referred to as a "one-light" print. They picked one exposure setting for the entire film. If your entire film was uniformly underexposed, then the lab could bring up the light level for the whole film. If your film was uniformly overexposed, they could darken the print a bit. If your film was exposed spot on, they could match that as well. But if your film had variant light levels scene to scene, once again, you were out of luck.

Sound films added yet another complication to the printing process. The 7399 was a silent stock. Some labs just made prints and transferring the sound from original to print necessitated bringing a second lab into the mix. That said, many labs did handle the sound transfer in house. After generating your print, they would post-stripe your film by adhering a sound stripe to your film after it was printed. They would then transfer the sound from the camera original to the print. It was possible to get good quality sound on your film this way, but generally speaking, the quality of the post-stripe wasn't as good as the camera original stripe manufactured by Kodak. What this meant is that often times the sound quality on your print would degenerate. Many Super 8 prints featured a murky, bassy, warbly soundtrack. Don't get me wrong, you could get good sound quality on your prints, but that too was a crapshoot.

For all of these reasons, a Super 8 print was rarely as beautiful or sounded as good as the camera original. As a filmmaker, you had to let that desire for perfection go, and accept degradation as part of the print process. In short, getting prints was a real hit or miss prospect. I got prints back that looked great and had horrific sound. I got prints back that had pristine sound, but looked terrible. I even got lucky upon occasion and got several prints with both nice picture and sound. It could be done! If you got a solid print, this was the ideal solution for exhibition, not to mention preserving the integrity of the camera original. However, for many filmmakers the frustration of dealing with the printing process led to exhibiting the camera original over and over and over.

Archivist Ross Lipman sums up the print conundrum nicely. In 1996, he wrote, "One would think that there would be a film printing stock intended to make high-quality copies from a positive image, but this is sadly not the case. Kodak's current 'internegative' stock, 7272, from which positive viewing copies can be struck, is a historical relic which actually greatly increases the final image contrast. The 7385, a low-contrast positive stock intended for 'TV prints' is sometimes used in conjunction with it, but this produces a desaturated look inappropriate for many subjects. And 7399, the reversal printing stock intended for this purpose, can no longer be printed with a high-quality soundtrack." So there you have it. Pick your poison. ■

CONSUMER PRODUCTS

Reduction Prints

ONE OF THE THRIVING Super 8 markets was reduction prints of Hollywood features and short subjects for home viewing. Before video, this was how you could own, watch and rewatch your favorite films. Most films were released in digest form on 50-foot, 200-foot or 400-foot reels. This means you were buying an edited version of the film running roughly three minutes, ten minutes, or twenty minutes. Many films were released silently, featuring newly minted title cards in lieu of sound. One of my prized possessions is a 200-foot black-and-white silent version of *Planet of the Apes* (1968). Clearly this was not what the director intended, but in the pre-video days this delivery sated some sort of need.

There were plenty of reduction prints that included sound, and believe it or not, there were also feature length films delivered on Super 8. The titles available ran the gamut, but if you were on top of your game, you could be projecting feature-length versions of *Jaws, Saturday Night Fever, Tommy, Dark Star, They Shoot Horses Don't They,* and *Straw Dogs* in your home theater. And people did set up Super 8 home theaters for just this purpose. Steve Osborne published *The Reel Image*, a magazine dedicated to Super 8 collectors, and he transformed his garage into a Super 8 cinema. It was pretty tight.

You could find all sorts of films on Super 8. Silent era comedies were popular titles. Chaplin films, Keaton films, W. C. Fields films, Abbot and Costello films, and Laurel and Hardy films are still easy to find. People loved their monster movies, as well as animated films from Disney, Hanna Barbera, and Terry Toons. You better believe sporting event highlights found their way to Super 8. Who wouldn't want to watch Ali and Spinks duke it out on Super 8? Better yet, what about Ali and Frazier going toe-to-toe in the "Thrilla in Manila"?

The choices of Hollywood films were vast. Did you want to watch Christopher Lee in *Gorgon*? If so, you were in luck. Columbia's 8mm division had you covered. Peter Falk in *Machine Gun McCain*? Check. Joan Crawford in *Strait-Jacket*? You got it. The quality of these prints was a mixed bag. Many of the prints were generated from 16mm copies, rather than the original 35mm negative. Reporting on this phenomenon in relation to a 1975 Super 8 print of Alfred Hitchcock's *Murder*, released by Thunderbrid Films, Anthony Slide notes, "[They] used a 16mm print from a 16mm negative, duped from a 35mm print, as a preprint, which means that the 8mm prints you buy will be some six stages or more removed from the original 35mm negative." Still, you could watch *Murder* at home, so there's that. If a major studio, like Columbia, MGM, or Universal was releasing its titles, there was a greater chance the digest prints would be "direct descendants of original negatives in the company vaults." ∎

> "Did you want to watch Christopher Lee in *Gorgon*? Peter Falk in *Machine Gun McCain*? Joan Crawford in *Strait-Jacket*? If so, you were in luck."

CHAPTER 10:

VIDEO TRANSFER

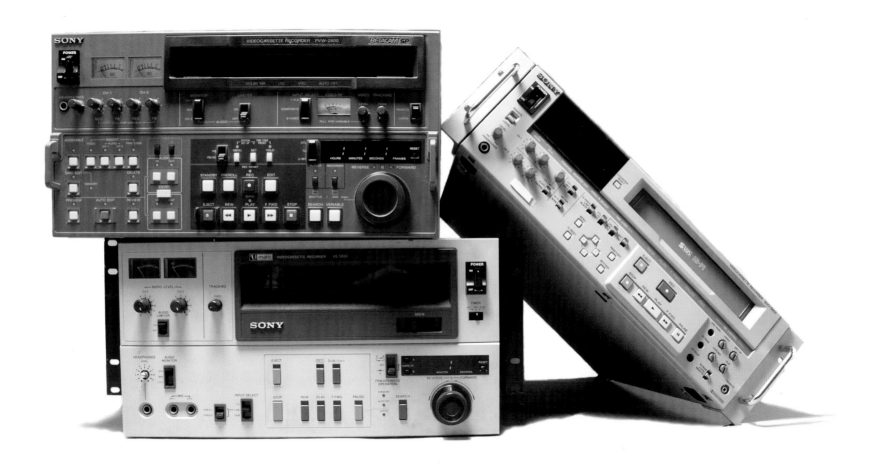

VIDEO TRANSFER

Though video exhibition was not a standard option on the festival and microcinema circuit until the early to mid-1990s, getting a quality video transfer was still a necessary endeavor.

Most festivals accepted video transfers as part of the submission process. If you didn't have a video, you would need to send prints or camera originals. That was not a desirable option, because it tied up key assets, not for screenings, but for potential screenings. Sending video allowed you to keep your camera original safe and keep your prints available for actual screenings. Transfers were also useful to send to press contacts, cable access television, and friends and family.

It should be noted that when you transferred your film to video, your film's image quality changed. Transferring to video gave you the ability to tweak colors and exposure. That's good. However, if you got a cheap transfer, the results could be a loss in dynamic range or color saturation. That's bad. Once you were in the video realm, you could then spend time equalizing your sound, adding sound effects, or creating a more complicated sound mix if you wanted. That's good. But it all cost money. That's bad. The range of possibilities available to you was closely linked to how much money you were willing to spend in the transfer process.

Before you made your transfer, you had several technical and financial considerations to come to grips with.

Film Chain vs. Rank Cintel

There were two broad types of transfers: the Film Chain and the Rank Cintel. The Film Chain featured a film projector and a video camera each pointed toward a multiplexer. The film image would be projected into the multiplexer where it would bounce off a series mirrors, ultimately being captured by the camera. The projector was outfitted with a five-bladed shutter, which was used to help eliminate flicker that would occur when making the conversion from Super 8's 18 or 24 frames per second frame rate to the 29.97 frame rate of video. The more money you spent on the transfer, the better the quality of the equipment, the operator, and the final transfer. On higher end systems, you could "time" your transfer, moving through the film scene by scene and adjusting your exposure and color levels accordingly. A "timed" transfer was superior to a "one-light," where you set one exposure level for your entire film. A "one-light" would prove to be problematic if your film had variant light levels.

Film Chains were the most common type of system. There were many dedicated Super 8 labs using film chains and producing very nice results with this setup. Film Chains were also the systems prevalent with services whose sole purpose was to transfer home movies. These operations tended to be the most economical way to go, charging in the neighborhood of $10/roll, plus the cost of video stock. Though they were economical, the results left something to be desired, as they tended to do "one-light" transfers with very mixed results.

Rank Cintel was a system where your film frame was scanned line by line by a flying spot of light. Working with a Rank Cintel operator, you would go through your film shot by shot, the operator making exposure and color adjustments along the way. As you worked, the operator would enter those timing parameters into a computer. Once the changes had been entered, the film would be played back, the computer making adjustments as the transfer proceeded. These systems tended to have greater control of light and color than Film Chain systems.

Companies with higher end systems, be it Film Chain or Rank Cintel, tended to charge by the length of their work time,

Film Chain

not the length of your film. These services ranged in price from $100–$300/hour. If you had a lot of light level changes, and the need for a lot of exposure timing, they might work on your film for five to ten times the length of your finished film. Transferring a ten-minute film in this fashion could run you upward of $300.

Format

Though VHS was the format of choice for home use and festival preview, it was never the transfer format of choice. VHS was the lowest quality of the tape formats. The goal was to transfer to the highest quality format you could afford, using that source as a master tape from which you would generate VHS dubs for distribution and festival preview. In the analog world, every time you made a dub, there would be image degradation as you moved from one format to another. In other words, when you made a dub from your master to VHS, the quality of your film got worse. Not only was there generational loss, but you were also dubbing to an inferior format. You know how your 1080 film looks when YouTube is playing it back at 240? Something like that was going on as your film made its way down the ladder toward its VHS landing spot. I guarantee that most filmmakers who came of age in the 1980s have no nostalgia for VHS. Sure, it allowed people to see your work, but you couldn't help but cringe when you saw your film degenerate.

Here are some of the popular transfer formats that were under consideration.

Beta SP was the king of the heap in the 1990s. The drawback of Beta was that if you wanted to do any post work in video, like clean up edits, or add a more complex soundtrack, you needed to go to a high-end studio to edit Beta. These studios cost upward of $100/hour and might also necessitate paying a tape operator to run the equipment.

1" tape was a popular format in the 1980s, but just like Beta SP, you needed to go to a production house to edit it. On a personal note, most of my early work was transferred to 1". This proved to be unfortunate because the rise of Beta SP meant the death of 1". Most production houses junked their 1" machines by the late 1990s, forcing me to dump all my footage to Beta. I never did have a knack for picking formats with longevity.

3/4" tape, sometimes called U-Matic, was a popular production and transfer format in the 1970s and 1980s. As you read through the interviews in this book, you'll consistently hear the likes of Dave Markey, Derek Jarman, and James Nares talk about entrusting the transfer process to U-Matic tape. The quality wasn't nearly as nice as 1" or Beta, but many media arts centers had 3/4" editing systems that you could rent for $15 or $20/hour. Most film schools in the 1980s also had 3/4" editing systems. Most university students learned to operate these systems, so there's a good chance that you or a friend not only had access to a 3/4" system, but also knew how to operate it. If you wanted to do some post-work in video to spruce up your Super 8, this was a good choice.

So, like everything in the film world, the more you spent, the better your transfers and dubs looked. When I started making films in the 1980s, I transferred to 3/4". It was affordable, and I could do some minor tweaking. Having access to 3/4" systems meant that I could also make a bunch of VHS dubs on my own. In the 1990s, once I realized my films had legs, I stepped up and retransferred them to a better format. Once I committed to 1", I could no longer generate dubs on my own, and became reliant on dubbing houses for my duplication needs.

Off the Wall

With the rise of MiniDV in the late 1990s, many filmmakers would just choose to transfer their films off the wall. You'd project your film on the wall, set up your MiniDV camera and hit record. The results were certainly a mixed bag. Not only were you essentially making a "one-light" transfer of your film, but there would be some keystoning as well, since the camera and projector were at a slight angle to each other. You might also get some video roll bars cruising through your picture as well. But it was quick, and it was cheap, and for home movies, or rolls that were of lesser importance, this guerilla tactic often did the trick.

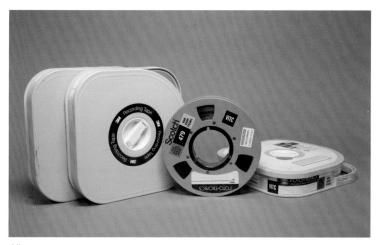
1" tapes

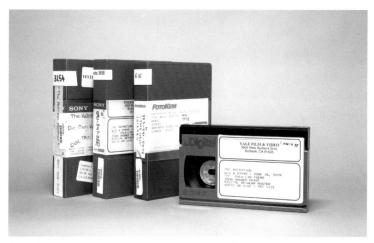
Beta SP tapes

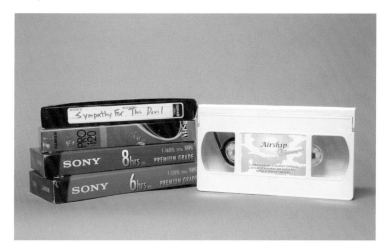
VHS tapes

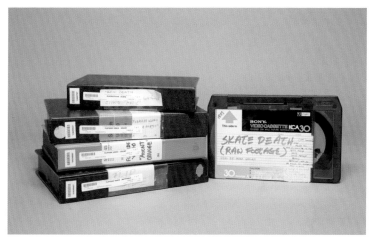
3/4" tapes

Super 8 to 4K and Beyond

Today, film transfer technology is reaching a new level. Many of the artists in this book are beginning to transfer their Super 8 films to 2K and 4K. There are top of the line film scanning systems like the Kinetta 5K scanner that is the transfer system of choice for archivists. These systems are able to go back in to the Super 8 originals and pull information out of the films in a formidable way. Filmmakers can then use high-end color grading systems to further enhance their films. James Mackay, who has overseen the transfer of Derek Jarman's films, puts it this way: "What Derek was seeing through the viewfinder is what we see now with advances in scanning. We see the clarity of picture that he saw, rather than the projection which had not caught up with the technology of Super 8." This is a crucial point. At its core, the Super 8 image contained a lot of information. Most projector technology couldn't deliver that image quality, once the image was blown up large during projection. Now with 4K scanning, we can begin to get a glimpse of the image the way it was shot and meant to be seen. ∎

CASE STUDY

I'm Not Fascinating, The Transfer! A Nightmare In Many Parts

TRANSFERRING WAS OFTEN expensive and challenging, but transferring my 1996 film, *I'm Not Fascinating,* was a nightmare beyond compare. At the time, I had a transfer house I loved. They did great work, were great people, and they cut me deals. They were based in a different part of the country than I was, and normally I sent them my films with notes for the transfer. They transferred the films without my supervision and they always did top-notch work.

I'm Not Fascinating was a fifty-minute sync sound film, and was filled with many visual blemishes. Given the scope of the film and its visual challenges, Anthony Bedard, my coproducer, and I traveled to the East Coast to supervise the transfer process (something I had never done). It was a fantastic experience, and we even got to see Halley's Comet during the sessions. The transfer process went swimmingly with one minor hiccup. The sound head on the film chain was out for repair while we were monitoring the sessions, so we transferred the picture only, and the transfer house was going to take care of the sound the following week.

When we got the finished film back, there were half a dozen scenes out of sync. When you transfer Super 8 films to video, there's no question that certain sacrifices have to be made in terms of picture quality but, in my eyes, sync was nonnegotiable.

It was one of those tricky situations because the transfer house cut us a great deal, and put in tons of extra hours they didn't charge us for. They gave us a great looking film and they were hospitable beyond belief. But all of that said, I had to call them up and tell them we couldn't accept the transfer. Things got a bit ugly and we had to move on. This interaction remains one of the major regrets in my film career. People in the Super 8 world were imbued with a DIY ethic, continually championing each other's works and services. Having to go through an ugly negotiation with people I had worked with for years, and had a tremendous professional respect for, pains me to this day.

After my print fiasco with *PIPSQUEAK PfOLLIES* (see page 94), I decided not to attempt a print of *I'm Not Fascinating*. This decision made the necessity of generating a high-end video transfer even more important.

We screened our camera original at our premiere in February of 1996, as well as during an East Coast tour in March of 1996. Over the course of the spring and early summer, we had a spate of shows up and down the West Coast, all of which featured our camera original Super 8.

We decided to put off the second transfer attempt until July when I could supervise the transfer at a reputable transfer house closer to home. I was very clear with them regarding the scope of the project and the type of transfer that needed to be done. We had an edited Super 8 sync sound film, and we needed a 1" transfer of it. It seemed pretty cut-and-dried to me.

The transfer house primarily transferred Super 8 for advertising agencies. Most of their clients were getting transfers of dailies that would then be edited at a later date in different video or film formats. Because the house primarily had to concern themselves with transferring individual shots, as opposed to completed scenes, their Rank Cintel system was not programmed to accommodate a completed, edited film.

Their plan for our film was to individually transfer each shot, presenting us with eighty-plus shots, which would leave us with the unenviable task of having to recut the entire film on video at an expensive

118

EQUIPMENT

editing house. We didn't want to do that and convinced them to program their system to accommodate our edited project. We rescheduled for a month later.

We drove down to LA for what was now our third attempt to transfer the film. Things seemed to be going smoothly. We went through the film scene by scene, entered the timing info into the computer, and then went to make the final transfer. We sat back, ready for the machine to handle its business. As the transfer started rolling, things got weird. At each edit point the film started shifting around wildly. Major registration problems were occurring at every single edit. With Super 8 transfers, you could expect a little bit of registration shift at the edits, but this was off the charts. A quick examination revealed that the tension on the machine was so great that the film was stretching to the breaking point, each tape splice pulled to the extreme, leaving visible gaps between each cut. The film couldn't be transferred in this condition. Once again, though the lab was great and the people were professional, we found ourselves stymied by technical limitations.

We were dejected, to say the least. That afternoon, we returned to where we were staying and were pretty bummed. We were sleeping on the floor at the house of a friend who ran a small record label. When we got back to his house, he was on his way to the Beastie Boys' G-Son studios to gather up a couple indie rock superstars and head to the Chateau Marmont for drinks. Did we want to go? It was a Hollywood dream night in the offing, but we weren't worthy. We asked for directions to the closest bar to drown our sorrows and had a couple of stiff ones before taking a six-hour drive of shame back to Northern California. Once again without a transfer of the film.

Back in San Francisco, I had to resplice the entire film before returning to the lab for one more crack at the transfer. The final transfer took place without incident. We were beaten and bruised, but at least we finally had a transfer. C'est la guerre. ∎

The Supermatic Video Player

KODAK WAS CHASING the professional and TV news markets with the one-two punch of their Supermatic 8 Processor (see page 101) and their Supermatic Video Player, which were released in 1974 and 1975 respectively. The Video Player was a countertop transfer system that allowed you to display your film cartridges on TV. It operated on the flying spot scanner setup that was similar to what was under the hood of the Rank Cintel system. The pitch was that if your company or station owned both the Processor and Player, you could quickly develop your own film and then transfer it to video. The Video Player, with a price tag of $1,350, was a steal compared to the five-digit sticker price of the processor. It also gave us an incredible ad campaign. ∎

On the set of I'm Not Fascinating!

CHAPTER 11:

EXHIBITION

EXHIBITION

Before YouTube, exhibiting shorts was a difficult proposition.

There were very few outlets for short works. A handful of cable channels might purchase a short, and European television was a distant option, but for the most part filmmakers turned their attention toward the festival circuit.

There were a few dedicated Super 8-only festivals such as the Ann Arbor 8mm Film Festival, US Super 8 Film & Video Fest (Rutgers), Splice This! (Toronto), Super Super 8 (San Diego), Brussels Super 8 Film & Video Festival, and VIVA 8 (London). A greater number of festivals accepted Super 8 alongside 16mm, 35mm, and video. Those with a strong Super 8 component included Athens International Film Festival (Ohio), Onion City Film Festival (Chicago), Montreal Festival du Jeune Cinema, Micro Cine Fest (Baltimore), and many others. Global Super 8 Days also sprung up in many cities. Alas, due to projection issues, there were many festivals where Super 8 simply wasn't invited.

Once video projection became more prevalent, opportunities opened up for Super 8 filmmakers. With a decent transfer in hand, you could submit a Super 8 film to a festival with the understanding that it would be screened on video. Also, the 1990s brought with it a revitalization of underground, no-budget, and punk cinema. This was reflected in the exhibition world by the emergence of an underground film festival circuit. The New York Underground Film Festival was the first in the ring kicking things off in 1994. It was quickly followed by Chicago, Boston, Miami, and Honolulu, to name a few. All of these were very receptive to Super 8 low-budget missives.

There was also a small circuit of cinematheques and microcinemas that took a chance on Super 8.

Media Arts Centers also played a crucial role in helping emerging filmmakers realize their visions. These organizations were invaluable to artists at a time when the modes of production could be cost-prohibitive, and exhibition opportunities were few. They were places where artists came together to share ideas, plot, scheme, and be inspired. Media Arts Centers served many roles. They rented equipment at bargain-basement prices. They offered classes so filmmakers could learn the craft. They screened the work of their constituents, offering open screenings, salons, and local festivals. Their presence was an essential cog in the development of underground and independent film scenes worldwide. They were spaces where people making non-mainstream work could band together to produce impassioned films. They were homes to filmmakers taking chances.

From the 1970s to the 1990s these Media Arts Centers dotted the landscape. Film Arts Foundation and Artists' Television Access in San Francisco, 911 Media Arts Center in Seattle, Squeaky Wheel Media Arts Center in Buffalo, Pittsburgh Filmmakers, London Filmmakers, and The Funnel in Toronto are a few of the landmark institutions. With the rise of digital production and distribution in the 2000s, many of these institutions folded up shop, though several still remain to fight the good fight.

A less rarified, but equally exciting, exhibition space was Public Access TV. Cable networks were required to provide stations on their systems dedicated to government, educational, and public programming. These stations aired local, community produced content. Often there was an underground movie show on these channels. Getting your film screened on Public Access was usually no more difficult than sending them a U-Matic tape. In the 1980s and 1990s, media thrill-seekers spent many a late night

viewing bizarre oddities from across the country courtesy of Public Access.

Finally, more resourceful and daring filmmakers took to touring. Super 8 was ideal for that because the projectors were so compact. You could put your projector in your car and roll. A handful of underground artists, using the model of punk rock touring, took to the road exhibiting their films and those of their friends. They pieced together screenings at more conventional film spaces like cinematheques, media arts centers, and college film programs, while simultaneously filling in the gaps of a touring schedule with events at off-beat cafés and music venues that were willing to take a chance on an evening of underground/punk rock film. Melinda Stone took her Super Super 8 Festival on the road in this manner, playing anywhere from laundromats to film festivals. I did a lot of this type of touring in the 1990s. A typical day on the road might feature an afternoon guest speaking engagement at the local college, followed by an early evening workshop on Super 8 production at a media arts center, followed by a midnight screening at a microcinema.

As the 1990s progressed and VCRs became a living room staple, filmmakers could also market their titles on home video. This necessitated pulling together a ninety-minute, feature-length program and finding a distributor willing to take a chance. There were a handful of distributors who put out underground work, often distributing to record stores and edgier, metropolitan video stores. Additionally, by identifying niche markets and figuring out where to advertise, you could also sell tapes via mail order to people in more rural areas who might not have access to a cool record or video store. ∎

ESSAY

Festival Blues

UNTIL THE ADVENT of the internet and streaming services like YouTube, filmmakers who lived on the fringes were reliant on festivals to exhibit their work. Not surprisingly, festivals could be a blessing or a curse.

One of the oddities of the festival circuit was that it was a pay-to-play universe. You paid to submit your films. And you were more likely to be rejected than accepted. Submissions tended to outpace available screening slots. If you had a 50 percent acceptance rate, you were doing pretty darn good as a filmmaker. And the cost of submissions piled up. Application fees cost anywhere from $10-$50. You then had to factor in shipping, the cost of your VHS submission tape that wouldn't be returned to you, the cost of a bubble-pak mailer, and the cost of copying the one sheet you would submit with the film. You might even have had to mail a self-addressed stamped envelope to receive the festival application form in the first place. How we functioned before the internet was mind-boggling!

As mentioned earlier, one of my first festival frustrations was around the handling of *Skate Witches*. To recap—I was rejected by a prestigious, local 8mm festival who then handed out the local filmmaking prize to an interloper from out-of-state. Clearly they thought my film sucked. A half-million internet views later, I get the last laugh, but thirty years ago I was nursing my wounds.

My subsequent film, *Dumbass from Dundas*, was embraced by the festival, receiving an honorable mention. It was my first festival accolade and I was pretty stoked. The next year, I submitted my follow-up, *Death Sled II: Steel Belted Romeos*. I had high hopes. In my estimation, *Steel Belted* far surpassed *Dumbass* in terms of storytelling and technical sophistication. I won't say I was dreaming of top honors, but I expected to be in the running for something.

Frustration slowly began to seep in. Festivals had submission dates several months prior to the festival, allowing a selection committee to view the films and decide what to screen. If your film was selected for screening, you would be contacted and asked to send a film print. It was fairly standard practice to submit your film on videotape, and send the actual film once you had been accepted.

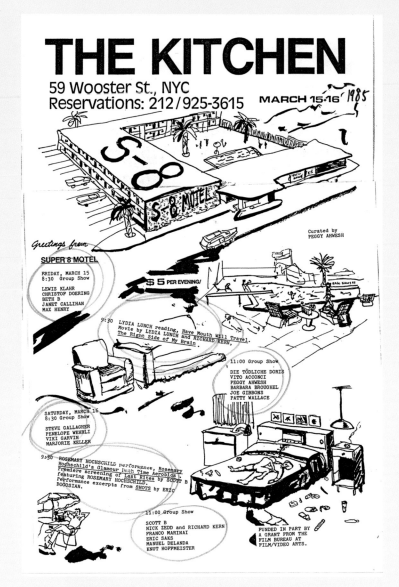

Super 8 Motel flyer, March 1985.

Otherwise, you'd need countless prints to submit to numerous festivals. That wasn't a sustainable practice.

As the festival approached, I had received no word that my film was accepted. Three weeks out, I phoned the festival to check on the status of my film. Radio silence. Two weeks out, I called again. Radio silence. At that point, I had resigned myself to the fact that the film had been rejected. On top of that, I talked to several friends who had received acceptances. Disappointed doesn't begin to cover it. I made one more pestering call a week before the festival. Radio silence. I was pissed.

Then a strange thing happened. The morning after the festival ended I received a call from Jim Sikora, a good friend, and one of the best Super 8 filmmakers out there. At the time, I lived in San Francisco, so the film festival was no longer local to me and I wasn't in attendance. He excitedly let me know that his film had received an honorable mention in the experimental category. Tamping down my jealousy, I enthusiastically congratulated him. Then he said, "You'll never believe what film won top prize. Yours did!" I was flabbergasted. How did I win top prize? I hadn't even been accepted. What version of the film did they show? Why had they never contacted me or returned my calls? I won, but somehow I was miffed.

I hung up and immediately called the festival. This time they actually picked up. Yes, I had won. Yes, they had shown the VHS submission copy. That was a bit disappointing. At the time, I only had mediocre transfers of my films and the VHS dupe from the 3/4" master was not that appealing of a screening format. But, whatever, I won. My prize? A brand new, high-end video transfer of my film. The bitter irony of course was that to get my brand new, high-end transfer of my film, I had to send them my camera original film!

Wouldn't it have been nice if they already had it in their possession, which would have meant they would have screened the film on Super 8?!

I was reluctant to send the original film. *Steel Belted Romeos* was actually a film for which I had managed to secure two quality prints, so I sent them one of the prints. Ideally you would send the camera original for the transfer, but again, I felt there was too much risk involved. As it turns out, I made the right choice. My prize package turned out to be a curse or a sick joke. I sent the print off, and the festival stopped returning my calls. A month went by and I hadn't received my print or the transfer. I called to see what was up. No response. I kept calling. Two months. Three months. Four months. Same deal. No response. They had my print, they hadn't made the transfer, and they weren't returning my calls. This was maddening. I only had two prints of my film, and one of them was AWOL. Running out of options and patience, I had a lawyer draft up a threatening letter. No response. It seemed to be a lost cause. Many months later, I had planned a visit to my hometown. I decided to make one last-ditch effort and called the festival, letting them know I would be in town and I was coming to get my print. I couldn't care less about the transfer, but I needed that print back. They responded. We made a plan. I showed up at the office and they weren't there. Classic. I called again, left an angry message, told them where I was staying in town and to drop it off on my porch. They did. Crisis averted.

Granted this is somewhat of an insane story, but it also does a great job encapsulating the festival experience. It costs a lot of money, there's a lot of rejection, and sometimes there are projection problems! ∎

ESSAY

Life on the Road: Small Town Blues

ONE OF MY FAVORITE screening stories involves being contacted by a high school student who had seen my work at a summer arts program. As part of a senior project, he was tasked with putting on an event for his local community. He lived in Colfax, California, a small town in the Sierra foothills that didn't even have a real movie theater. This was the type of screening that was exciting to me. I was heading off to a small town to screen my work to a group of people who never had the opportunity to see underground films. That said, this was the type of screening full of potential pitfalls for a travelling artist. The gig was two-and-a-half hours from home and my travel time would necessitate missing a day of work. Additionally, small towns were a tricky proposition when it came to attendance. Sometimes you could get a great turnout, simply because there was nothing to do in such a place. Sometimes, however, no one would show up, because no one lived there. As payment I was being offered half of the door. But you know the saying, "Half of nothing is nothing." The high school student was convinced at least a hundred people would show up (since attendance was vaguely required for his graduating class). I was leery of that promise, but I decided to roll the dice. I gave the high school student one task to help ensure that he was invested in the screening being a success. I told him he had to provide a screen.

I drove on up, and when I arrived I instantly had doubts. This town was really small! As I wandered Main Street, all one block of it, I had the uneasy feeling that there was no way a crowd could be

ESSAY

The Gay Festival of Super 8 Films

THE GAY FILM FESTIVAL of Super 8 was founded in 1977 in San Francisco by a group of filmmakers including Marc Huestis and photographer and filmmaker Danny Nicoletta. The festival eventually morphed into Frameline, now the oldest LGBTQ+ film festival in the world. In the mid-1970s, Huestis picked up a camera at City College of San Francisco and began making Super 8 films. His films would win awards at festivals like the Ann Arbor 8mm Film Festival and the International Festival of the New Super-8 Cinema in Caracas. Huestis recollects the early days of the Gay Film Festival of Super-8 Films in this excerpt from his book, *Impresario of Castro Street: An Intimate Showbiz Memoir.*

NOW A PROLIFIC FILMMAKER, I needed to buy and develop vast amounts of Super 8 film. The only place to go was a half-block away from my apartment: Castro Camera, owned by Harvey Milk and his lover Scott Smith. There was little doubt that Harvey was a mover and shaker, and the atmosphere of political activity inside the Castro Camera drew you in.

Manning the front desk at Castro Camera was an adorable young waif named Danny Nicoletta. While Danny and I chatted at the front desk during my visits, other Super 8 gay filmmakers would drop off films for processing and join in our discussions on filmmaking. Eventually, we decided to support each other in creating a public exhibition for our fledgling film projects. We called it the Gay Film Festival of Super 8 Films.

summoned in this sleepy town. I arrived at the auditorium and my mind was simultaneously blown and put to ease. The student and his friends were in the middle of the auditorium floor, bare feet, surrounded by power tools, and they were building a screen. I assumed that he would be able to borrow a screen from school, or find one at a Salvation Army. Not so. He had decided to undertake building a screen from scratch. With help from his friends, he made a massive screen. He had researched screen materials, and when I arrived they were in the process of affixing the screen to the sturdy wood frame. It was awesome. At the end of the day, it was a great screening. There was a nice-sized crowd, approximately fifty folks. The beauty of this type of small town screening was that it was a pretty interesting cross-section of the population. It wasn't just urban hipsters. It was moms, dads, families, and high school kids.

At the end of the night, the high school student sheepishly apologized for not pulling in the a hundred folks he promised. Embarrassed, he handed over all the door money. He had promised me 50 percent of a hundred admissions. There were only fifty folks, and he felt obligated to give me the entire take. It was one of the sweetest offerings I had ever witnessed on the art circuit. I refused and insisted on taking only the 50 percent of the door that we had agreed upon. I had never believed there would be a hundred people there anyway. And the kid and his friends totally delivered. As we hung out in the parking lot after the show, I could sense they didn't want the evening to end. I suggested that we grab a bite at a diner or a Denny's, whatever the town had to offer. They were so excited until they realized there was no place open after 9:00 p.m. to grab a bite in town. So off I headed to San Francisco, a two-and-a-half-hour drive in front of me. ■

The author in the projection booth at The Artists' Television Access, San Francisco.

ESSAY

Our ragtag group of hippies, nerds, and filmmakers included Danny, Bern Boyle, Ric Mears, Wayne Smolen, Billy Miggens, Greg Gonzales, and David Gonzales. We formed a loose collective and met in my Castro Street flat. I'd like to claim that I cofounded the festival out of some altruistic, utopian vision of a burgeoning new gay culture. But really, darling, I just wanted my ego stroked and my damn movies to show. And there was safety and power in numbers.

Our fledgling festival was loosely curated. Show up to a meeting and your film was most likely in. With one glaring exception. Rob Epstein's submission, which featured his lover John Wright naked in a bathtub, intercut with shots of his cat cleaning herself. The group decided his short was "not gay enough" and rejected it. Of course Rob would eventually go on to win two Oscars for his groundbreaking documentaries: 1984's *The Times of Harvey Milk* and 1989's *Common Threads*.

My life was now dedicated to organizing the festival, writing press releases on a battered manual typewriter and copying them on a dilapidated Xerox machine. Silvana Nova designed the fire-engine-red poster and Neighborhood Arts donated the printing. Soon every telephone or streetlight pole on Castro Street advertised the festival.

The festival had a distinctive Mickey and Judy flavor. Instead of a barn, however, we used a funky community space at 32 Page Street, run by gay activist Hank Wilson. Danny Nicoletta contributed a rickety Super 8 projector. We couldn't afford to rent a screen, so we hung a funky white bedsheet. I even ironed it!

On February 9, 1977, the festival was held. Admission was free. Two hundred people arrived for a space that could only accommodate a hundred. The overflow was turned away. (Except for Harvey Milk, whom we ushered in.) As pot smoke perfumed the air, we ran the films, coping with the numerous film splices that broke and the audiocassette soundtracks going out of sync. The audience didn't care; they were there to see images of the lesbian and gay community, something rarely addressed by the Hollywood machine.

Our maiden voyage spawned three other screening events that year and would later evolve into Frameline, now the oldest and largest LGBT international festival in the world. We had created something special. ■

CHAPTER 12:

DEATH, REINVENTION, RESSURRECTION, RE-EVOLUTION

DEATH, REINVENTION, RESURRECTION, RE-EVOLUTION

I made my first Super 8 film in 1985, and it seems like Super 8 has been dying ever since.

By that point almost no companies were making new cameras. Likewise, new projectors had also stopped rolling off the assembly lines in the US, Germany, and Japan. You could still find all the gear used, but with each passing day that equipment inched closer to death. Rubber capstans decayed, battery contacts corroded, light meters showed signs of cataracts, and motors wheezed to a halt on these thrift store finds. Favorite cameras died, others were kept alive by the good graces of tinkerers and hobbyists, resurrecting ailing cameras, allowing life to breathe into one more 50-foot cartridge.

But one by one, they picked off my favorite film stocks. My first love was 4-X. It was killed in 1990. Super 8 sound stocks were the heart and soul of my film universe. They bit the dust in 1996. Kodachrome, the richly hued fan favorite, got sent to the abattoir in 2005. For many, this felt like the final curtain. The prized jewel in the Super 8 stable was no more. Plus-X was quietly shown the door in 2010.

Ektachrome A & G were killed off in 1996, but like a Terminator, Ektachrome never stayed dead for long. It rose from the ashes in 1997 only to be killed again in 2004. It popped back to life in 2005 only to perish once more in 2010. The process repeated again, with another resurrection in 2010 followed by another death in 2012. But lo-and-behold, Ektachrome is back with a brand new lease on life in 2018, joining Tri-X, the black-and-white workhorse, as the sole survivors from Super 8's glory days.

If your goal is to shoot Super 8, cut Super 8, and project Super 8, then this is indeed an accurate history. But remember, Super 8 is a slippery beast, continually being reimagined by those drawn to its flicker. Super 8 is forever morphing, its history shaped and reshaped by those attracted to the magnetism of this small gauge format.

My history revolved around making Kodachrome sound films. When pre-striped sound films were eliminated, the way I made films was significantly impacted. But that's just the way I made films.

For those who shroud themselves in a more experimental framework, for those not concerned with sync sound, myriad silent stocks and production possibilities still exist.

For those willing to shoot on negative stocks, for those with access to double system recording techniques, and for those willing to finish their films digitally, a new realm of hybrid possibilities abound.

Nowadays, with digital distribution and exhibition so prevalent, why not work this way? In the digital realm, Super 8 films can be cut with much greater sophistication than they could in the Super 8 heyday of the 1970s, 1980s, and 1990s. Straight cuts and simple sound mixes, which were a limitation in the old days, can be a thing of the past.

When I teach Super 8 to my college students, they love it. They love the look and the feel. The Super 8 palette is unique, wholly different from the HD look to which they've grown accustomed. Though it's old technology, Super 8 is a new tool in their tool belts. A new color on their color wheels. I make them shoot reversal. We go old school. Not because that's what I love, but because I know the thrill they'll get when they see their film projected for the first time. It's a singular experience.

It's not often that I see Super 8 projected anymore, but when I do, I fall in love all over again. I'll be the first to admit that I love the beauty and the clarity of high res, HD images. HD has been trying to look like film for years, and it gets closer and closer every day, maybe having already bridged that gap. But every time I see small gauge film projected, the difference is palpable. You can see it. You can feel it. This isn't nostalgia talking. These are your eyes drinking in the images. The way the grain dances across the screen is a thing of beauty.

I'm glad that Super 8 remains alive. For a format that has been dying for thirty years, more than half its life, Super 8 remains strangely resilient.

And hey, Kodak plans on releasing a new Super 8 camera. That camera is inspired by the designs of a Danish father and son team, who produced a small batch, high-end Super 8 camera in 2013. I love that this reinvention is a father and son endeavor. For a format whose claim to fame was dominating the home movie world, that development seems sweetly appropriate.

Over the course of its fifty-year existence, Super 8 has always meant different things to different people, and that remains the case today. You can still find cameras at flea markets, in thrift stores, and on eBay. You can still shoot reversal, cut on film, and exhibit on film. You can play your film silently, perform a live score, or accompany your film with music played out of your iPod, phone, computer, tape deck, or record player. You can shoot negative, digitize your film and edit with the postproduction software of your choice. You can create elaborate soundtracks, layer image upon image, and mix your Super 8 with HD images, 16mm images, or any image of your choosing.

There's no right or wrong way to make your film. How I made films in the 1980s may have been foreign to Lenny Lipton's method of production in the 1970s. How some young artist, head filled with fiery images and incendiary stories, makes a Super 8 film today may be foreign to the way I did it. In fact, I hope it is radically different. If it were the same, I'd be disappointed, and, worse yet, bored!

So, what are you waiting for? Pick up a camera and make your own damn movie! ∎

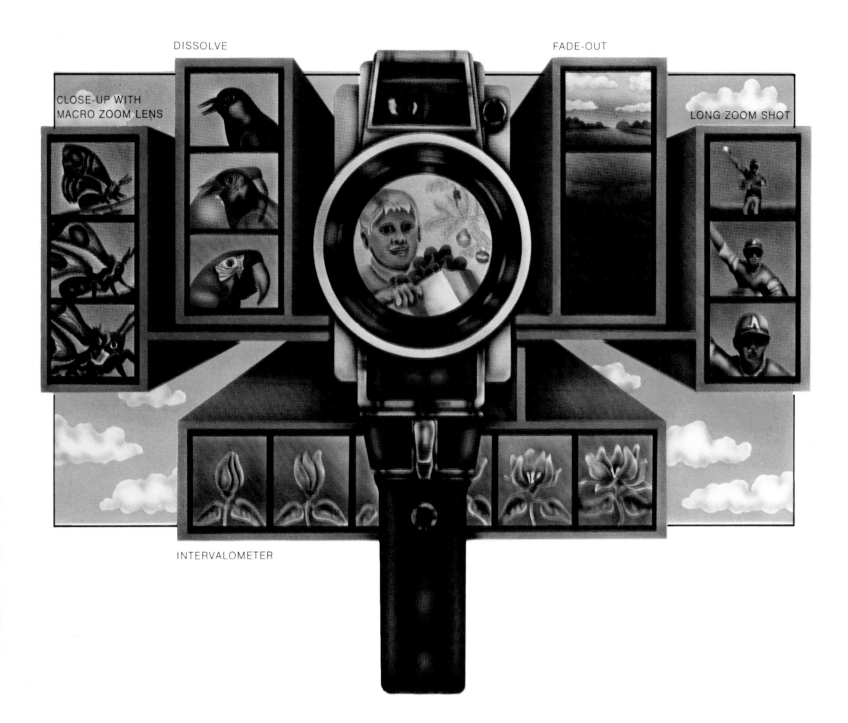

INTERVIEWS

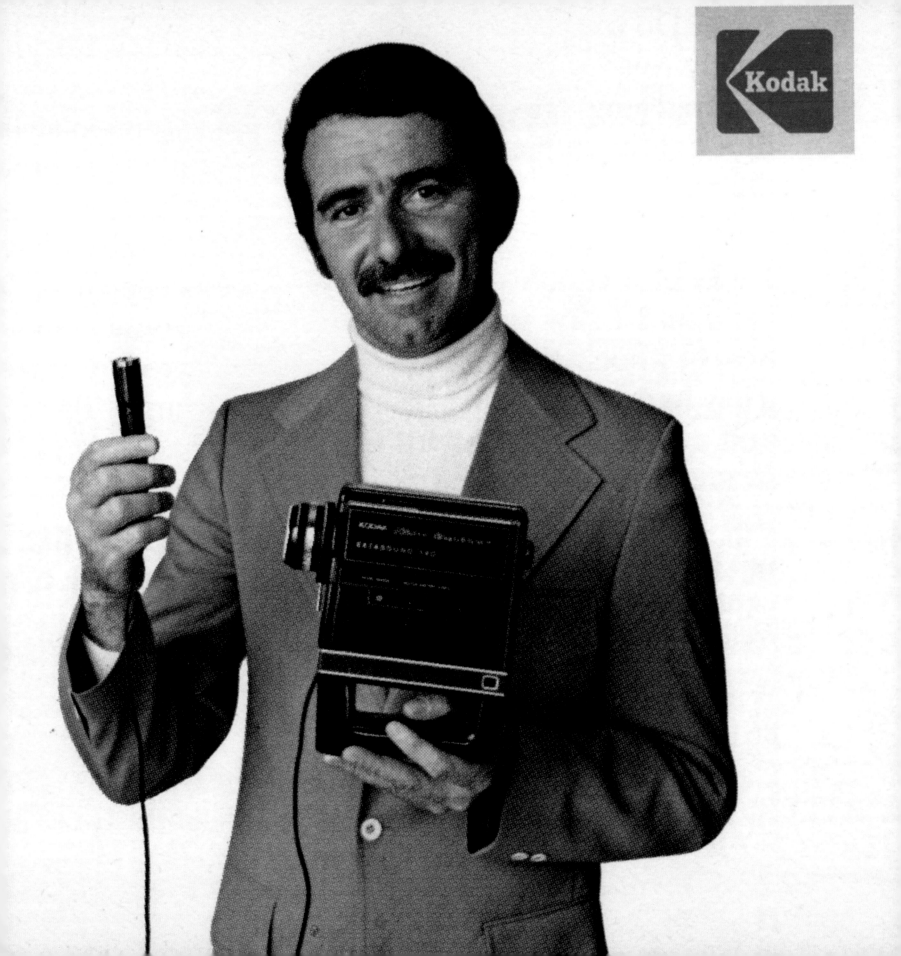

THE 1960S AND 1970S

The 1960s era of Super 8 was really defined by home movies.

A new, easy-to-use, low-cost tool was made available to families as a way to document their lives. Sales went through the roof.

At the outset of the 1970s, Super 8 was still trying to define what it was from both an aesthetic and technological standpoint. The possibilities were endless. Certainly, it was being used by kids to do what kids do—make silly films starring their friends, mining the repository of popular culture for inspiration. Famously, in suburban Detroit, Sam Raimi (*Evil Dead II, Spider-Man*) and friends were engaged in making low-rent horror films. In Los Angeles, a young Dave Markey was creating epics Like *The Movie of Movies,* his fourteen-year-old take on *Kentucky Fried Movie*.

By the mid-1970s, college-aged folks were starting to get serious. Rocky Schenk was making Maya Deren-influenced dreamscapes. John Porter was documenting public rituals, embracing the limitation of the 50-foot reel. In New Orleans, Vance DeGeneres and Walter Williams turned to Super 8 animation, making the hapless Play-doh character Mr. Bill, who rocketed to fame courtesy of *Saturday Night Live*. In New York, Beth and Scott B were using the immediacy of Super 8 to make filmic broadsides whose plotlines were pulled from the headlines of the day. With the advent of single system sound, a new path for both narrative and documentary presented itself. With a mic plugged directly into his camera, Lenny Lipton began creating portraits of the artists populating the Bay Area subcultural movements.

But Super 8 was still seen as less than. In the opening of Lipton's *Children of the Golden West*, from 1975, we see Lipton approach 16mm filmmaker Herb De Grasse. While De Grasse futzes with his 16mm gear, bemoaning a missing instructional manual, Lipton proselytizes about the possibilities of Super 8 single system sound.

Lipton: Tell me, what have you got against 18 fps?

De Grasse: Nothing.

Lipton: Suppose you were just making movies and showing them in Super 8? Wouldn't that be okay?

De Grasse: Well, you can't do A/B rolls in Super 8.

Lipton: So what?

De Grasse: Well, that's a relevant thing at this particular stage.

Lipton: For you.

De Grasse: Well, if you could do Super 8 stuff, like you can do 16, it might be worth it.

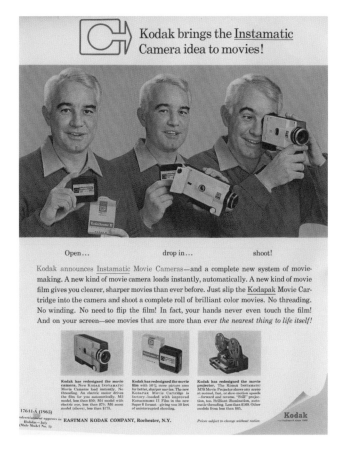

Lipton: Well, I'll share this to you when it comes back. I like to know what you think of how it turned out.

"It might be worth it," says De Grasse, who doesn't seem wholly convinced by Lipton's argument for Super 8's efficacy. This skepticism was not uncommon. Super 8 was fighting an uphill battle in its bid to be taken seriously. Gerry Fialka, director of the Ann Arbor 8mm Film Festival from 1977–1980, suggests that 8mm was perceived as "minor league" compared to the "major league" treatment that 16mm films and filmmakers received in the same time period. Regardless of perception, quality work was being produced in Super 8. Scouring the 1970's festival programs from the Ann Arbor 8mm festival, you find names like Todd Haynes (*Velvet Goldmine*, *Carol*) and Alex Gibney (*Taxi to the Darkside*, *Going Clear*). Rocky Schenk also showed regularly at Ann Arbor; his works were some of the most visually accomplished films in the medium. British experimental filmmaker Derek Jarman (*Sebastiane*, *The Last of England*) made stunning Super 8 work throughout the 1970s.

Not only could Super 8 craft a beautiful image, it also offered a glimpse into the world of young people. Jarman, Lipton, and many others, such as New York-based filmmaker James Nares and Argentianian filmmaker Narcisa Hirsch, documented the people, the spaces, and the ideas that populated their scenes. They created an intimate view into the world of a new generation of artists and ideas, a viewpoint not regularly documented by popular media at the time.

One of the interesting threads running through many of the conversations with filmmakers from the 1970s was the challenge of finding your people. Derek Jarman and James Nares, both born in England, shared many of the same touchstones, yet they never saw each other's work. John Porter wandered the streets of Toronto looking for his people with little success, and he was a letter carrier! Many filmmakers making great work didn't know about the 8mm festivals that were cropping up worldwide, or even that a Super 8 magazine, *Super-8 Filmaker*, was on the newsstands. Artists often worked in isolation, or with small groups of people, not always aware of the larger movement out there.

The potential of Super 8 was also exploding on the technological front. Kodak's introduction of low-light film cameras in 1971 was a big deal, allowing filmmakers to shoot with available light. The advent of single system sound in 1973 was a game changer. There was a bit of a space race to create double system sound, which would allow Super 8 to function like 16mm. The pages of *Super-8 Filmaker* functioned as a guide book to all matters Super 8 and were filled with all the latest doodads and gadgets to satisfy the tech nerds out there.

The following batch of interviews give great insight into the lives of artists working with a film format that presented its own set of challenges and triumphs, and for which the rules were still being written.

Please Note: Some interviews and questions have been slightly rearranged and edited for clarity. ■

INTERVIEW

Roland Zavada

IN 1965, WHEN SUPER 8 was developed, Kodak was one of the biggest companies in America. They attracted chemical, mechanical, and optical engineers from the country's top universities. Roland Zavada was one such individual. He first became interested in photography while developing Verichrome film with his father in the basement of their home in Gary, Indiana. Zavada's first camera was a 127 that he purchased with cereal box tops. With World War II in full force, he enlisted in the Army Air Corps. Not yet eighteen, Zavada enrolled as a chemistry major at Purdue, where his interest in photography led him to join the school yearbook staff. In July of 1945, his class was called up for military service, and Zavada headed to Biloxi, Mississippi, for basic training. However, with the war ending, he was given the choice to return home or continue his tour in the Air Force. He chose the Air Force and volunteered for aerial photography gunnery school. He was shipped to Okinawa, where his background in photography and chemistry landed him the job of lab technician in the first photo reconnaissance squadron, which was focused on map making. By 1946, the squadron was flying reconnaissance missions over China in a B-29, taking aerial photographs of the burgeoning communist country. After his military tour, Zavada landed at Kodak in 1951, where he became a product engineer specializing in reversal film technology. While at Kodak he worked on teams that redesigned Tri-X and Plus-X, and introduced Kodachrome II. Zavada was a principal player on the team that designed the new Super 8 motion picture format. He was on the ground floor, grappling with the technical, chemical, and engineering challenges of bringing Super 8 to market. In 1966, he was tasked with the oversight for the national and international standardization of the new Super 8 system. He was the Chairman of the Committee on 16mm and 8mm Technology for SMPTE, the Society of Motion Picture and Television Engineers, and eventually served as its Engineering Vice President from 1974–1982. Zavada retired from Kodak in 1990 as their Standards Director for Imaging Technology.

By reading the initial 1964 SMPTE article by Edwards and Chandler that introduces Super 8, it seems like one of the big reasons behind developing Super 8 was for educational and industrial markets.
Oh, yeah, that was the goal. The philosophy was that [Standard] 8mm was simple because it essentially was split 16mm with just twice as many perforations and it worked effectively. But in terms of image area it was not efficient. We said, if we could improve 8mm to where it's almost 16mm, if you're in the right environment, you have that. I was the first guy to project 8mm with a twelve-foot width. In cooperation with General Electric, we developed a halogen bulb. We recognized that a two blade shutter [for a projector] was inadequate because of heat at the film plane. So we developed a three-blade shutter and that's how we came to the 18-frame-per-second pulldown to be able to get that in terms of three-blade shutter effectiveness.

What were some of the other challenges in developing Super 8? I know that getting proper perforation size and placement was a big issue. There were also challenges with the cartridges, correct?
You're going with a smaller perforation. You have to be able to perforate such that you're not going to have residual emulsion hanging on, which is a dirt factor. And you have to have enough energy to be able to move the film without distorting it. That's where we came to the space gate concept.

[Editor's note: One of the key developments of Super 8 technology was including the pressure pad (or plate) in the Super 8 cartridge. The pad was critical in keeping the film in focus. In 16mm and Regular 8, this device is part of the camera. The inclusion of this device in the cartridge was one of the factors that allowed Kodak to develop their easy-to-load Super 8 cartridges. When Zavada refers to the "space gate," this is what he's talking about.]

When you start thinking of the Super 8 cartridge and what we call a space gate, you had to develop a certain film curl so that when it was held down on the edges you had a relatively flat exposure surface. The physical properties became a very important consideration in the development of those films. So what you do is have the film curl [a certain] way and

INTERVIEW

the space gate holds it. You have a pressure plate in the cartridge not in the camera, and that was the concept that allowed us to have a drop in cartridge. Twenty second reload capability. And that was very important because that's where friction came into being. That came into mass production molding of the plate. You'd be amazed at the hell of the problems that we could get in. The guys in the plastic industry didn't quite realize that their temperature stability was critical to a flat gate because there was no machining afterward. It was mold and use.

I read in the Home Movies book that 300,000 test rolls were shot while developing Super 8. What were you looking for?
I don't remember 300,000 rolls! One of the things that I did is that I was a giveaway guy. I'd say, "Here's three rolls of film. You shoot them this week and bring them back to me. I'll develop them for free, but I get to examine them." That was a thing that we did to see how people would work, especially when we came out with the movie camera that you could hold this way. [Editor's note: Zavada is referring to the Kodak XL camera.] People wanted to [hold it the wrong] way. We had to look at a lot of film to say, "How do you develop the instruction manual? Don't let them make mistakes that they'll then blame you for."

Super 8 takes off in the home-movie world. Was that anticipated?
We knew that there was going to be big amateur opportunities. One of the big voids that we were trying to fulfill was the educational medium. Essentially we were a little late with that by a few years. There were not adequate judgments in terms of what was going to happen in the videotape industry. The guys in research never thought videotape would be successful so dynamically.

In your 1970 article, The Standardization of Super 8, you name standardization as a key issue. With Super 8 registration, is the issue at the camera level or the projector level?
At the printer level. The camera level is going to give you a position. Now you want to have cost effective educational [prints]. With Super 8 you're going to print three rolls of film on 35. The thing is you can only guide one of those three prints. That's it. So it becomes critical for the laboratory to do their lineup properly. If they get one film right, and, damn it, don't examine the adjacent films for steadiness…so we had to develop all sorts of steadiness tests and methods of evaluation and that was a lot of work.

Early on you're also doing magnetic sound on the educational films, right?
Magnetic sound was by application, and that was a big challenge to be able to apply a thin stripe with enough flux density in it to be able to carry reasonable sound.

You'll find us in every port. And at booth 33.

Sound prints happen right away, but sound cameras aren't developed for another eight years.
It's difficult to say why we didn't develop a sound cartridge concurrently with the silent. We had a block of wood, knowing what size our camera people were going to have to work with. But the picture/sound separation, we weren't absolutely sure where it was going to be, because now we were trying to develop high-quality, low-cost projectors. And that was a big challenge for the department. The thing is that the projector is a heat generator, and you're in a situation that is contrary to effective film transport.

The other thing I found interesting is that Kodak shared their plans for Super 8 with their competitors. Is that standard or not standard?
No. The publication that came out first was by Evan Edwards and Jasper Chandler (SMPTE, Volume 73, July 1964). That was the one talking about a new format film and that was because we wanted to get standardization. A lot of people say, "Do standards follow practice?" Yes, they do, if you have good prepractice. So we went to all the manufacturers and said, "This is what we're doing. This is what we have developed. Here are the results of this transport technology, which you can incorporate in a camera if you adopt the format." Then the format became acceptable, and with that we were able to standardize.

So you went to Bell & Howell and others?
All that. We were able to do it. But Japan was developing a Single 8 system.

Were they developing it concurrently?
It was hard to say. I interviewed the director of the Film Division of Fuji and he said that Japan was very conscious of the need for visuals in their education system. They

INTERVIEW

were looking at the fact that Standard 8 was an inefficient format and they wanted to improve it 50 percent. So they were developing their Single 8 system. The other thing is Kodak—we had geniuses who could develop three-part molds. The Super 8 cartridge is a complex mold. That's great, except if somebody else wants to use that technology. It is so costly for them to make cartridges that they were looking for simpler ways, and [Single 8] was like reel-to-reel film. Kodak's desire of having a simple cartridge including the gate was different than Single 8, which the gate was in the camera. So if you're a manufacturer saying, "Look, I can achieve the same thing, I'll have the same format and I can make the camera gate more precise than a space gate," how do you get the guy away from that? You can't by saying convenience is worth it. But they were cooperative.

With Kodak letting their competitors know about the new format, where do you think they're making money? Stock and processing? Camera sales?
When you're making your money, you're going to have to do it in film sales. So you're looking for something that's going to allow for compatible, instead of like Fuji. In the role I had in standards, I go to a meeting in Williamsburg and my company is expecting me to come back with a Super 8 cartridge standard. I have everybody from the world over there, and I'm chairing the whole damn thing. And Japan comes in with Single 8. So how do I get a single standard? What did

I do? I held up the standard on the Kodak cartridge. The question is how many raises did I lose because I did that? And while Japan developed, I took all of the specifications for the Single 8 and I drew up both standards. So now I said, "Okay, here we are. Instead of one standard, you have two. You have approach A and approach B." I didn't like that. That was not a happy part of my career.

Ultimately it seems like the Kodak version…
Is the one that won out. Why? The thing is, you have to load the camera. Your situation is you're a youngster or something and you just [mimics popping the cartridge into the camera]. Make sure the batteries are adequate and you're ready to shoot.

It's probably easier for the manufacturer and less expensive to develop the camera with a space gate, right?
From the camera standpoint, yes. All you need to do is have the pulldown claw and you have to have the plate-to-film distance. That becomes your critical factor in the gate.

Another thing about standardization that you talk about in the article was the sound displacement issue. The article predates single system, so what's going on there? Why wouldn't people adopt a standard?
What is one of the critical things when you record? Stability, consistency, uniform motion. What do you do with film at the gate? It's intermittent motion. So you have to have time between the intermittent motion and the sound recording to stabilize that piece of film to consistent velocity. That's where you develop the picture/sound separation standard. Now you test and test. You want it as close as you can get it because otherwise you look at the sound cartridge and the sound cartridge is elongated and still only has fifty feet of film in it. But it had to have the space for the sound head and it drives the camera manufacturers anxious because they don't know what you're going to come up with. And they're engineering a two- to three-year lead time. It was a challenge.

INTERVIEW

Lenny Lipton

What was the initial reaction when people started seeing Super 8?

You've got to remember that it was not a single point of the system. You got the Super 8 format, then you had a projector that was able to display that format, and then you had a new lighting system which was able to display an image that was brighter and bigger than you had before. So the impact was phenomenal because it was a combination of the fact that you had a larger image area and you were able to utilize that in projection. It was great.

The situation was that at Kodak we had other factors that we had never done before. You've seen the racks of processing machines for motion pictures. They never put an 8mm through the tension of those rollers in the tank and successfully got it out the other end as the same piece that you put in at the beginning. We had to develop soft touch rollers. We had to develop rollers that had a spring in there so that if you put pressure on the roller, you didn't affect the whole rack of eight rollers. It affected that one roller. That was a fantastic development.

One of the other big challenges with Super 8 was getting prints from reversal film. Did Super 8 offer new challenges or were those the same challenges?

Super 8 followed the initial development of Kodachrome II. I was a product engineer on that introduction from the coordination standpoint. Not the emulsion, but coordinating manufacturing. We were able to handle and develop a Kodachrome duplicating film. We know the curve shape and we were able to put the curves so that we could reproduce that. We ended up, from an 8mm standpoint, looking at this thing on a forty-inch screen at home. The difference between the original and a duplicate was entirely in your acceptance range. You're not saying, "Oh, I'm looking at something too contrasty or hard or jumping all over the screen." You could make for your wedding guest, everybody can get a roll of film of the wedding, and they all thought it was the original.

Were you involved with the development of 4-X?

We developed it as a high speed film. I think that was going to be a bank surveillance film in a Super 8 cartridge. The bank could record the events that were taking place at their counter as a single frame system and they would store the film cold. If they had a problem they could go back and develop it to see what was on it. We knew the banks had a problem and they needed something. So we developed that 4-X film and a Super 8 cartridge with an intermittent motion surveillance camera.

One thing that I read was that Super 8 was always split from 16 film, not from the 35. Is there a reason for that?

Well, 8mm came from 16 film by being half of it. Super 8 changed that from the standpoint that now 16mm could be used as the originating medium for audiovisual technology and the aspect ratio was now fully compatible that it'd be a simple reduction.

Is there anything else?

Let me show you my Beaulieu camera. [Ed note: At this juncture, Roland shows me his Beaulieu 4008 ZMII camera and asks if it is something my school can use.] Why don't I give it to you?

You don't need to do that?

I don't. But realistically, I'm ninety-one and I'm not going to be doing any filming and I don't necessarily just like to put trophies on the shelf.

[Editor's note: When one of the original designers of Super 8 offers up a top shelf camera for you to take home, you can't say no, can you?] ■

IF THERE IS ONE MAN whose name is synonymous with Super 8, it's Lenny Lipton. After all, he literally wrote the book on Super 8. In 1975, Lipton published *The Super 8 Book*, an essential read for any young filmmaker ready to take the plunge. Prior to the arrival of Super 8, Lipton made 16mm documentary films and worked as an editor at *Popular Photography*. This position gave him access to the new Super 8 equipment as it came to market. Once Kodak introduced single system sound recording, Lipton was a full convert. *The Super 8 Book* is part how-to manual and part consumer report, written with a hint of evangelical fervor. Lipton believed in Super 8 as a tool to democratize filmmaking. His film, *Children of the Golden West,* opens with a number of conversations about the potential of the

INTERVIEW

medium. Talking to Lipton, filmmaker Paul Sawyer says, "The communication is now open to people with smaller means to make maybe thitrty minute films at something like $300, $400, $500 at the most…So the situation is wide open. There's no excuse now for entry of revolutionary imagery into the American consciousness…It's not just smoke. It's real." Lipton's voice extended beyond the pages of The Super 8 Book. He remained a frequent contributor to Popular Photography, and throughout the 1970s you could find his writings in the pages of Super-8 Filmaker, the nation's sole magazine dedicated to Super 8. He reported on new gear and traveled to festivals to report on new films. Well into the 1980s, The Super 8 Book remained an important tome. Lipton was also an accomplished filmmaker, one of the few to make Super 8 films in 3-D.

What were you doing before the advent of Super 8?
In 1965 I was working in 16mm. I did shoot in 8mm too. I tried different styles but mostly I wound up making glorified home movies that some people thought were of some ethnographic importance. I lived in Berkeley during the flowering of the countercultural movement, so I used my films to record my adventures of what was happening. You could look at them as documentaries. I couldn't afford 16mm synchronized sound—it was too expensive. I lived in the Bay Area. I usually did not use collaborators and I was stuck on the idea of being a filmmaker the way a painter or a writer would do their work. I thought that I would express what I was feeling and what I was seeing without using helpers. I wanted to shoot it and I wanted to edit it and write it and put it together myself.

Were you originally from the Bay Area?
No. I worked in New York City for Popular Photography magazine, as the editor of the movie section. I got married and my wife was a student at Berkeley so I went to Berkeley with her. I was one of the original members of Canyon Cinema. I was on the board of directors. You know there's a very active independent or underground or experimental filmmaking movement in the Bay Area. It was very stimulating. It was fun. There were many interesting makers there.

So you are not collaborating, but who were your kindred spirits?
The filmmakers who I knew at the time like Larry Jordan or Will Hindle, Robert Nelson, Ben Van Meter, John Schofill, Herb De Grasse, Richard Broughton. I knew all these people and everybody had a different style. So it wasn't a movement in the sense that the activists were. It was not a unified movement in terms of style and I'm not sure anybody else was making films quite the way I was. They were all doing things a different way. Robert Nelson was involved in something called the Funk Movement.

Lenny Lipton shoots 3-D Super 8.

James Broughton and Bruce Baille were both using film as poetic expression. I mean I did too, but I think the bulk of what I did was as I described.

Kodak announces this new medium in 1965. What was your response and what was the general response in the community? Were you somewhat guarded or was Super 8 met with enthusiasm from the get-go?
When Kodak announced Super 8 I was an editor of Popular Photography magazine and I was assigned the story. So I covered Kodak's introduction. Kodak divulged their plans in advance to their competitors for some kind of nominal licensing fee. So Bell & Howell was taking a very strong position introducing Super 8 equipment and I covered that too. I visited Bell & Howell in Chicago. I visited Kodak a number of times. I had access to Super 8 equipment. I met the people who developed it and got very

INTERVIEW

interested in Super 8. I shot a lot of Super 8, but I didn't start to use it for my own filmmaking until Kodak introduced their sound cartridge, which was just a great invention. The 200-foot version was really wonderful. It was great and I was able to shoot films with that. I got a grant from the American Film Institute. I made a film about a guy named Cirengiva Roy, who was a guru who lived on Scott Street in San Francisco. He had thousands of followers who followed his religious precepts. So I made a one-hour documentary about him. That thing is pretty good. Today the ability to take the camera film and transfer it to digital files just provides a missing piece. Maybe too late because you couldn't really make very good Super 8 prints by any method. Kodak tried to come up with an optical printer and so forth. But you know the problem with celluloid cinema is that you lose a tremendous amount in printmaking. Once you start making prints, especially if you're involved in making a large number of copies for theatrical release, especially in color, you move away from the original intention pretty fast. But scanning films to digital files and then releasing it as a DCP, that really hits everything that's in the original reversal or negative. It's a way to scoop up all that data and get a clean look. That was a big problem with Super 8, and it's been solved too late.

In your book, you are definitely a champion of single system sound and tape splicing. But there definitely seems to be a whole group of people working double system and doing cement splicing in the seventies.

It doesn't matter, cement splice or tape splice. The problem is that I had worked with work print in 16mm and, you know, working with the original is really the shits. I think I turned it into a virtue. I mean what the hell are you going to do? Making work prints would've made it more costly, and as time went by I just really didn't have the money and I just decided that I would do the best I could by cutting the original. But it's impossible to make changes in some circumstances. There are all kinds of things for single system sound like displacement recorders or you could just record the sound onto a Super 8mm mag stripe and cut double system. And I did those things and got pretty good at it. I used interlocked Eumig projectors. I had a great time in my studio because after years of waiting, I was able to do sound. The thing about single system sound is that if you're a one-man band, it's really much easier. Once again, extending the idea of not using a crew, it's just you, and if you're shooting conversations in a documentary like that, people will look at you and pay attention to you. You're less intrusive.

While reading your book, it's clear that you have access to so much gear. Was that because of your connections at *Popular Photography*?

Yeah. I mean I was reasonably personable and I kept writing for *Super-8 Filmaker*. I made many friends in the industry, at Kodak, and at other companies. I knew people at Paillard, in every company, and so I could borrow. And I have an eccentric personality that was pitched to a reader. I'm sure it was very engaging. Magazines have to have columnists who are characters, and people look forward to that. So I really did like writing that column and enjoyed Super 8 because it was a liberating movement all over the world, not just in the United States. There were people there in Iran and people in South American countries who were shooting Super 8. And those people were doing ethnographic studies of indigenous populations that were on the edge of disappearing. People in Venezuela or Brazil were going out, shooting tribes in Super 8. I was invited to festivals there and it was a really nice time. I was very lucky because the underground movement in the sixties was a nice fellowship. It was fun. And then I segued to the Super 8 movement, which was more of a worldwide movement. I really enjoyed it because it was part of a big community of people who were very friendly and we were all in it together. Kodak unleashed something. There's no reason why they couldn't have made a product like Super 8 twenty years before. They could have done it. It may sound like sour grapes. I mean Double 8mm was a terrible terrible idea. I know why they did it and it made a lot of good sense. They could use the 16mm labs. But the film was foggy and the possibility of running the film through twice was such an inefficient use of the film. The area of the frame was so small and the perforations were too big. It was bad design. Super 8 was a very good idea. Who knows what filmmaking would have been like if they only did it sooner. But you can always say things like that.

"Super 8 was a worldwide movement. I really enjoyed it because it was part of a big community of people who were very friendly and we were all in it together. Kodak unleashed something."

INTERVIEW

Do you know what the impetus was for making Super 8?
The impetus was mostly for industrial because they thought they could extend film distribution in those areas using Super 8.

You mean like using reduction prints down to Super 8?
Yeah. And it did happen. For years, airplane movies were shown on Super 8. You see, they couched it that way in their preliminary article in which they were talking about the format. I think they had to do that to avoid killing 8mm sales, which died immediately. But it was such a better format for the amateur home movie maker. So much simpler. But you know, they really never followed through with proper projectors. The projectors remained with 1895 technology. There are many things they could have done to make the projectors nicer. But it didn't happen. They were clunky. In order to really have succeeded, they would have needed to make quiet projectors. The technology for a continuous drive was there and they just didn't use it. I think that for the home movie maker and even for the classroom they had a wonderful opportunity to design quiet running machines and they didn't do it. There was well-established technology for optical image stabilization that would have worked beautifully for Super 8 and they didn't do it.

Do you think the bigger problem was projectors or not being able to get good prints? As you're talking about these great ethnographic films from South America and Iran, without good prints could you see them if you weren't at the festivals in those countries?
The answer is probably no. Although if you blew them up to 16mm you got a pretty good result. But you didn't get a good result on Super 8 to Super 8. It was mushy and the colors were off and everything. Mostly you were starting with Kodachrome which wasn't meant to be reproduced.

Lipton edits the right and left bands of a 3-D film on a 2-gang editing bench.

The other thing that I find fascinating with Super 8 is that because it's a home-movie medium, there are so many different cameras. Do you feel that helped or hurt?
It was great. Almost all of the cameras had built-in zoom lenses of some considerable capability, and you had low-light cameras with very wide shutters. You had cameras from Nizo and Bauer and Paillard. You had all kinds of designs. You even had Double Super 8 for a Bolex camera.

Were a lot of people using double system? In your book there's a huge section about this. But by the time I started making films in the 1980s, all that seemed to be gone.
People who were dedicated to making Super 8 really could cut double system sound and do a very good job of it with relatively inexpensive equipment. You could get two projectors, if they had inching knobs you could watch them together. There were Super 8 recorders that were modified from portable reel-to-reel recorders. There were Super 8 multi-gang synchronizers. That was all you needed. It wasn't that expensive and it worked. I don't know about the eighties. I think what happened, the steam went out of Super 8, not for everybody, but as soon as camcorders were introduced, Super 8 was crippled. I can't remember the date exactly, but by the mid-seventies it was pretty clear that videotape was going to fulfill the needs of most of the amateurs, which was the basis for the market. And they don't edit or worry about making prints or things you're talking about. With videotape, for far less money, you could run the tape endlessly and tape everything. And you don't have to project. You can just show it on your TV set. You

INTERVIEW

John Porter

didn't have to set up the screen, set up the projector, which is a big pain in the ass, or lower the room lights. For many reasons it was just a better idea and Super 8 couldn't compete.

When do you move away from Super 8?
I didn't. I made a bunch of films in Super 8 and then I got interested in stereoscopic filmmaking and I continued to use Super 8 for another five to six years doing experiments with stereoscopic filming in Super 8.

I've never seen any of that work. What years are these that you're doing that?
Well some people saw it and remember it vividly. I started in the early seventies and worked through the rest of the seventies. I made one film that I thought was pretty good, Uncle Bill and the Dredge Dwellers, and I went around and showed that in different parts of the country and other parts of the world. People would say to me it was so beautiful, it's so much more beautiful than they could do in the cinema. It was better than the 35.

Are there any stand out Super 8 films from that era that you remember or could name?
I saw films all over the world. I recall being impressed by the amount of effort that went into some Super 8 productions that seemed to be inordinate and loony. Like costume dramas in Venezuela of historical events. I'll never forget those. But I don't even remember the titles anymore.

Do you feel Super 8 fully realized its promise?
That's too vague for me. I just had a good experience. It did what it needed to do for a while. It could have gone on for quite a while without the introduction of electronic cinematography and videotape. It is what it is. I regret that Kodak didn't do it ten years earlier. ∎

JOHN PORTER'S PERSPECTIVE on the world of Super 8 is an important one. His career spans the entirety of the Super 8 timeline. Inspired by Lenny Lipton's film column in Popular Photography, Porter picked up a camera in the 1960s in Toronto and started shooting little narratives. He attended Ryerson Polytechnical where he found 16mm to be unnecessarily cumbersome, leading him to develop a philosophy around making films quickly using Super 8. During the 1970s, armed with a Nizo camera, Porter began making films in his Condensed Ritual series. These were short, silent films documenting cultural rituals like Toronto's Santa Claus Parade, wedding receptions and circus rehearsals. Porter fell in with The Funnel, a seminal experimental film community in Toronto. In the 1980s, utilizing Super 8's portability, Porter developed a type of live cinema performance he called "camera dances," which combined hand-held projectors and choreography to utilize the entire exhibition space as a screen. The 1980s also found Porter collaborating with the feminist, queercore band Fifth Column. Together, Porter and Fifth Column worked on a series of films to be projected over the band while they performed. (See G. B. Jones interview, page 187.) Porter continues to make films and teach Super 8 filmmaking workshops to this day. He still only shoots Super 8 reversal. Instead of being put off by the limitations of Super 8, he leans into them. Every time a film stock dies, Porter takes it as a challenge, and figures out how to keep creative.

Talk to me about picking up your first Super 8 camera. When did that happen?
I was serious about still photography in my teens, wanting to go to university to study photography and subscribing to Popular Photography and Modern Photography. In one of them, Lenny Lipton was writing a regular column. There was one that caught my interest about how your home movies don't have to be just a record of your family Christmases, vacations and birthdays. You can perform a little story. I had never thought of that. I loved that idea. My still photographs had been very cinematic. I was getting my friends to pose scenarios, and making photo books with the little narratives. After seeing Lenny's column, I went out and rented a Super 8 camera from this woman who was well known among film students for renting film equipment cheaply. I wrote a narrative inspired by a National Film Board [of Canada] short I had seen. It was an anti-war film using toy soldiers that was a critique of war toys.

What year is this?
The summer of 1968, my first year out of high school. I took some friends with costumes and props to the local vacant

INTERVIEW

lot to shoot. As soon as I started shooting, looking through the viewfinder and hearing the film run through the camera and realizing that I was recording moving photographs, I realized this is what I want to do for the rest of my life. Before I got the film back, I went out and bought a projector, a splicer, and a camera.

What did you go out and buy?
A DeJur which I've never seen since. It's very basic and a bit heavy.

Filmmaking is hard and not straight forward, how was picking up those skills?
This first film was very easy. It was just one roll. I did some editing after I got the film back. There was a technical problem with the splicer I bought. The sprocket keys were out of alignment in the splicer, so the splices weren't perfectly aligned, so it wouldn't go through the projector. For years I couldn't project it until I got the gumption to go back and redo every splice. I could only look at it on a viewer. Then I went to Ryerson. They had a photography and film production course. I was planning to study still photography, but by the time I got there I was more interested in filmmaking. It was all 16mm there and I didn't like that. I like to say one of the biggest things I learned at Ryerson was how I hated 16mm. All the equipment was expensive and big and heavy and you needed a crew with a Nagra recorder. It just seemed unnecessarily complicated. I made about twenty 16mm student films. I was still shooting a bit of Super 8 on my own time, but they wouldn't accept Super 8 films. That's where I developed my politics and philosophy that Super 8 is a better medium. It's better for expressiveness like a painter's brush or a poet's pen. You can be free with it. I developed this philosophy that you can make fine art with the simplest tools. That was evident from all the other fine art media. My goal was to make films as simply, cheaply,

> **"The less you have the more creative you have to be. I've always tried to impose that on myself. Limit what's available. I like to joke that I like it when they discontinue the stock. It forces me to be more creative."**

and quickly as possible. In other words, in-camera editing, no sound, one cartridge, three minutes, do everything yourself—the acting, the editing, the projecting. So you can make a film in half a day and show it as soon as it gets back from the lab. That's always been my philosophy. $50 film budgets.

Is there a reason the school wasn't accepting Super 8?
Ryerson was geared toward training people to get jobs in the film and the photography industry. Commercial art, commercial photography, and commercial filmmaking. The idea was you go there to get trained to get employed. I learned I didn't like the film industry. I thought it was excessive. I wanted to spend my whole life rejecting all of that, and trying to prove that you can make good films, serious films, fine art films on Super 8. When they wouldn't accept that, I said, "You're challenging me. I'm going to prove you wrong." But there was nobody working in Super 8 at that time, so I felt alone. I started my own little screening series called Autobiograph, from a combination of Edison's biograph and autobiography, which was another term for auteurism. So I was showing work that I could find for free from the National Film Board and from public libraries that lent out films. And I was a letter carrier, a postman, and it was on one of my routes that I met one of my ex Ryerson classmates who was involved with

The Funnel experimental film theatre. He told me about them and I started going to their open screenings and they were my community. That was when I discovered my people in Toronto. They had a real strident political position too, including that Super 8 was a serious format. All the other services in Toronto, The Toronto Filmmakers Co-op, The Canadian Filmmakers Distribution Center, The Art Gallery of Ontario, which was showing avant-garde film on occasion, it was all 16mm. The Funnel was sort of in response to all of these other services. We're going to treat 16mm and Super 8 and Regular 8mm equally. We're going to have production equipment in all the formats. We're going to publicly screen all these formats. Teach workshops in all these formats. Distribute all of these formats.

In between that period, there's a ten-year period where you're just kind of making your films in your own little world?
Even though Ryerson was a three-year course, I stayed for five years. When I realized I wasn't interested in getting a job in the industry, I wasn't interested in the three-year diploma. I paid no attention to the academics, so I kept failing every year. They'd try and kick me out, and I'd appeal and win the appeal and come back. I did that twice. Even though I wasn't interested in a diploma, I was still getting a great education and in university 50 percent of

INTERVIEW

your education is hanging out with fellow students, people your age with a similar interest and discussing things with them, and then seeing the experimental films being shown in class. So I hung out there from 1969 to 1974. I discovered The Funnel in 1978. They had been going for a year.

Were you aware of any kind of national Super 8 film scene like *Super-8 Filmmaker* magazine? The Ann Arbor 8mm festival?
No. It wasn't until I got to The Funnel, that I started to hear about Super 8 festivals happening around the world, and I was submitting to them, and getting shown.

Where were the various Super 8 festivals that you were submitting to?
There was one in Toronto started by the same people that started The Funnel. That festival started in 1976. Lenny Lipton came and showed his 3-D films two years in a row. I remember going to it but I didn't have anything that I thought was good enough to submit. It wasn't until I joined The Funnel, and inspired by them, that I thought [my films] were good [enough] that I started submitting.

Can you describe one or two of those earlier films that you're proud of?
I was interested in time exposure in my still photography. Shooting at night, taking long exposures and getting blurred movement. Then I heard about this camera, the Nizo, that does time exposures. I saw one for sale privately in the *Buy and Sell* newspaper. Ron Mann, a famous documentary filmmaker up here was selling it. He was sixteen years old and I was in my twenties. I had to go to the suburbs to negotiate with his mother to buy this camera. I still have the receipt.

How much did it cost?
Four hundred dollars, maybe less. It was an S 560. I realized in order to get time exposure you have to shoot single frame. Everything is sped up and it's automatically

John Porter "film busking" at Harbourfront, Toronto, 2001.

funny. I was challenged—how can I make a painterly film with these brushstrokes, blurred movements that isn't funny. So I made a pixelated film of a mother and child, a pixelated film of a landscape, trying to emulate classical painting. This is in 1976. It was these observational films that I did submit to the Super 8 festival and they were shown in 1977. As soon as I got that camera, I started shooting a lot of these public rituals. I call them *Porter's Condensed Rituals* because you see the whole Santa Claus Parade. I shoot the parade pixelated so you see the whole three-hour parade in three minutes. That's one of my best films that was shot in 1976. The *Porter's Condensed Rituals*, there's about 100 films in that series. *Drive-In Movie* was a document of a drive-in with three screens, and a projection booth building in the center. They would project in three different directions simultaneously, and they'd have double features, so they'd go through six movies in one evening. Two of the screens were minor screens. They were big but were sort of flat with reinforced stands. But this one big screen was hollow, and you could go inside and climb up this ladder to the top, and there'd be a four-foot-wide top lid with gravel on it. One of the things I enjoyed about the process of making these *Condensed Rituals* was getting a good vantage point from the top of a building. That's where I shot the Santa Claus Parade from. For the Santa Claus Parade, I shot from the top of a building where I was working as a machine operator at an insurance company. I got permission to go on top of their building on a Sunday afternoon. I got permission and I brought a friend with me for support in case I needed something. We hung out there with our coffee for eight hours. That was a whole discovery for me, learning things about these bizarre events, like people arrive at these drive-in movies two hours early, and there's a playground there and there are pinball machines inside the concession stand building. It's like a tailgate party before the movies start. I'd scouted this out in advance, and we needed to get up early and document everybody arriving and then the movie is showing on these two smaller screens, but I'm also seeing the cars directly below looking at the screen just below the camera. And it's a double feature. Between the two features people would move from one screen to another. After the last film was over I'd stay up there documenting all the cars leaving. When we arrived two hours before the screening, it's still light out. With lot of my films I would choose that time of day where you're seeing the sun go down and get dark.

In that period how are you editing?
All in-camera edited. In fact, the most extreme cases are the ones where it's

INTERVIEW

one shot. *Santa Claus Parade* and *Drive-In Movies* are essentially one shot, the mother and child film, the landscape films, just one composition. You just set up the camera on a tripod, start it running, when the film runs out your film is over.

What's the next phase of filmmaking for you?
When I started going to The Funnel and started seeing all these raw, anarchistic experimental films, that really inspired me. People were doing performance art, bizarre things. I realized that my *Condensed Rituals* were conservative by comparison. I had been interested in acting, so I got interested in combining performance with filmmaking. I started doing live performance with my projections, but also performing in my films. I'd call those camera dances. These are films where I'm doing choreography for the camera, and the camera movement is choreographed to create a unique dance that can only exist on the screen. Those are my favorite films because they are more out of control than my *Condensed Rituals*. *Centrifuge* was a really wild camera dance with the camera swinging around my head uncontrollably. *Scanning* was inspired by two different filmmakers and particular works of theirs at The Funnel. I combined what they were doing into this idea of shooting a film outside, panning around 360 degrees, turning the camera upside down, and shooting up into the sky. When I project it, I do exactly those movements with that small handheld Super 8 projector. I project around the cinema, around the wall, following the subject. It's a projector dance and the film is leading me. When people see that they are astounded. I took that idea out onto the street, handhold a projector, and it doesn't matter what I showed. In fact, I show my outtakes. It's good because it's changing all the time. Fifiteen seconds of this, and then ten seconds of that. People walking by are not going stay and see a three-minute film. I project on the sidewalk, on people walking by, especially if they're wearing white clothes. A white van is perfect. That's my film busking. The camera dances and the projector dances started in the late seventies and early eighties and the camera dances were made in the eighties.

Are you still making work today?
My peak period was the late seventies when I was making twenty films a year. In the nineties I started to slow down, partly because I was getting more screenings and I treat all of my screenings as a performance. I'm making one new film every two or three years now. The last one was a remake of *Scanning*. I made eight versions in the eighties. The one that I show all the time fell apart on screen. It was such a simple film to shoot, I always said, "If I lose this film, I can just go out and reshoot it," which I did.

No need for a print.
Well I don't like making prints anyway. Sort of contrary to my philosophy. That's why I like reversal. Filmmakers say, "Oh, you're showing your original, aren't you afraid it's going to get damaged?" I don't like that risk factor, but I also love showing my originals. They look better than any print.

Now you're down to Tri-X, so how do how does that affect your craft?
As I said, my philosophy was to keep things as simple as possible. Norman McLaren once said that he liked to limit his capabilities, his facilities, because it forces you to be creative. The less you have the more creative you have to be. I've always tried to impose that on myself. Limit what's available. I like to joke that I like it when they discontinue the stock. It forces me to be more creative. I imagine sometime in the future, and this is sort of dreaming, but fifty years from now there will only be one film stock and it'll be black-and-white Super 8 because it's the cheapest and the easiest to manufacture and to process, and I'll be happy with that. ∎

James Mackay on Derek Jarman

James Mackay

DEREK JARMAN MADE over seventy Super 8 films between 1972 and 1984. Simply put, his films are stunning, showcasing the beauty that could be wrested from the medium. Jarman got his start as a set designer on Ken Russell's *The Devils*, and directed a number of acclaimed features, including *Sebastiane*, a seminal entry in the LGBTQ film canon, and *Jubilee*, considered one of the all-time great documents of the British punk era. His Super 8 work traversed many styles; he documented his own working spaces and flats, and made beautiful landscape films and quasi-narratives. Jarman passed away in 1994, leaving behind an impressive body of work. His producer, James Mackay, has done a wonderful job transferring Jarman's films to the digital realm. Mackay also published

INTERVIEW

Derek Jarman Super 8, a beautiful book about the filmmaker's small gauge work. Mackay first crossed paths with Jarman at the London Filmmaker's Co-op, an important arts center devoted to avant-garde filmmakers, where he later became a programmer.

How did you end up in London with an interest in alternative film?
I went to London to study film at art school, North East London Polytechnic, at the end of '74. There's quite a strong course on film there, but not in the conventional sense. Because there were a lot of experimental filmmakers at the school, we had a strong connection with the London Film-Makers' Cooperative, which was very active at the time. We're at the high point of the structuralist, materialist phase that was represented by people like Peter Gidal, Malcolm Le Grice, Lis Rhodes, and others. Between the art school and the Film Co-op, I got my education. That took me through to the late part of the seventies, at which point I was running the cinema at the London Filmmakers' Cooperative, programming that. It was an interesting period because we had a lot of people coming through London then—Stan Brakhage, Paul Sharits, Danièle Huillet, Jean-Marie Straub. As part of my program at the London Filmmakers Co-operative, I asked Derek to screen some of his early Super 8 films, and that's how I met Derek.

In the *Derek Jarman Super 8* book you talk about how people at the Co-op weren't taking Super 8 very seriously, but you were.
Well my generation was. The Film Co-op had been set up around '66 and it developed into a 16mm workshop, which had its own printers and facilities. It was a workshop, cinema and distribution. It distributed films from member filmmakers in the UK, but also from filmmakers all across the world. Underground film, experimental film at that time was 16mm. In Britain it happened because television companies, newsreel gathering was on 16mm, so there were a number of small laboratories, processing and printing 16mm. Television companies would give dated stock to filmmakers cheaply. It was affordable, but that changed in the mid-seventies. Suddenly the companies switched over to electronic news gathering. Television used less 16mm and consequently [it] became more expensive. When I was a student in the mid-seventies, Super 8 was relatively cheap compared to 16mm. People I was at college with, John Maybury, John Smith, others, all were working on Super 8. But then the old guard thought that 16mm was serious, and Super 8 wasn't. So there was resistance when we refurbished the Co-op to spend money on high output Super 8 projectors, etc..

Can you talk about Jarman's path to making Super 8 films?
He worked with Ken Russell, designing *The Devils* and *Savage Messiah* so he was used to 35mm sets and he did make his first film on 16mm at the end of the sixties. He met [Ken's] wife on a train. I can't remember where they were going, but his wife was the costume designer on all Ken's films, and he met her and they got along. She introduced him to her husband, and he instantly gave him the job designing *The Devils*. Then a visiting architecture student from New York came to London and met with Derek. He had a Super 8 camera which he loaned Derek and that's how Derek started making Super 8 films. That's probably end of '71. For Derek it meant that it was under his control. The experience of working on the Ken Russell films made him a bit shy about working in the big studios with all that crew and bureaucracy. Derek preferred spontaneity, so he embraced Super 8 because it allowed him to do what he wanted. It's amazing how quickly he taught himself, how not just to shoot stuff properly, but how to edit it. The very first films were edited in-camera, but then he moves onto tape splicing. He very quickly moves on to cement slicing. Technically, he was quite proficient.

Why did he choose to go cement as opposed to tape splicing?
You see the tape splices when it goes through the gate. They overlap. The first ones he spliced used that system where it's not a straight cut, it does a squiggle; overlaps one frame into the other. That's not very satisfactory, and not stable. Then he's using the conventional tape splices, but they're always messy on Super 8 because unless the machine is perfectly aligned, you tend to get sticky edges and it's problematic. His cement splices, I mean even when we were scanning the material three or four years ago, it all held.

You talk about that first screening you did with Jarman where he was sort of live programming his films.
That's right. He had a canvas bag which contained his two Bolex projectors, an audio cassette player, and all the film reels

> "Derek preferred spontaneity, so he embraced Super 8 because it allowed him to do what he wanted."

INTERVIEW

that he was going to show that evening, and he would just pick a film at the time. He didn't have a preset program. He would judge the mood of the audience. He would play different music with the films. None of the films had soundtracks on them, and the soundtracks would change over time. He wouldn't always play the same music. But generally he had a thing that went with each one and it was on audio cassette.

Was that approach to exhibition radical?
I don't think it was radical for Super 8 filmmaking. There was a big thing in London at that time of expanded cinema, where people were doing stuff with multiple projectors. The fact that Derek was in the cinema with his projectors, changing the reels and talking as he was projecting the films wasn't that radical, but it was fairly unique to Derek. I'd seen other people do something similar, but Derek would continue talking even when the film started.

Do you remember which Bolex projectors?
It was the one with the different speeds 3, 6, 9, 12, 18. It did Super 8, Standard 8.

It seems like he had an interesting practice where he would shoot the films at slower speed but then project at even slower speeds.
The early films are shot at 18 frames. But when he got the Nizo camera and discovered that he could control the filming speed, and then he had the Bolex projectors that he could control the projection speed, he started to play around with different combinations of filming speed and different combinations of projection speed.

As an audience member what are the things about his films that drew you in?
Just the sheer beauty of them. I was totally immersed in the world of avant-garde and experimental cinema and they were unique and incredibly wonderful to look at.

He never seems to embrace single system sound. Is there a reason he didn't go down that path?
I don't think he was interested in synchronized sound. It would have stopped him playing around with speed. When we were working later on, like *Angelic Conversation, Imagining October*, he never said, "Oh, we should shoot this with sync sound." Even *The Last of England* hasn't got sync sound. I think it would have limited his ability to play with the image.

One thing that is interesting about his work is there are a variety of styles. I'm wondering if you can talk about his approach to those different types of films?
They start off by documenting his surroundings and people he knows. From that we then go to the more costume, more acted pieces. They overlap these phases, but there were things that are more people acting in front of the camera. *Journey to Avebury* is a landscape film. There's nobody in front of the cameras, it's just a place, as is *Studio Bankside* really. But then you get to things like *Miss Gabby Gets It Together* and films like *Tarot*, more structured, more narrative. But then he takes that away again and he starts to superimpose images from earlier films. He starts to build up layers of image with films like *Sulphur* and *In the Shadow of the Sun*, another phase that ends around '74. He then goes and makes *Sebastiane*, and then he embraces another style of filmmaking which is the continuous take, evident in films like *Picnic at Rae's* and *Gerald's Film*. He's not always doing the same thing. He finds another thing that works for him and he explores that until he's explored it fully, and then moves on. After *Sebastiane* there are less Super 8 films because he's making feature films, and they're obviously taking up more of his time.

Super 8 always had this reputation of being home movie-ish. When I look at Jarman's work I don't see that. I see fully realized films. Did he look at Super 8 in any kind of home movie or nostalgic way?
It wasn't home movie making. Derek used it as a way of making films without engaging with the politics of a crew. I think that's the real reason he picked Super 8. People see Super 8 as being a home movie medium because when you projected Super 8 using the Bolex projectors or whatever was on the market, you end up with a small, fairly fuzzy image on the screen. But cameras, especially the Nizos and the Beaulieus, were more sophisticated, and the pictures recorded on the film are much better than what you saw on the screen at that time. What Derek was seeing through the viewfinder is what we see now with advances in scanning and projection. We see the clarity of picture that he saw, rather than the projection, which had not caught up with the technology of Super 8.

As he moves into features what's that experience like for him?
The first three features, *Sebastiane*, *Jubilee* and *The Tempest* were shot on 16mm. So they were not quite as industrial as Ken Russell's films, but when we got to making *Angelic Conversation* it was closer to the smaller scale films that he had made in the early seventies. *Angelic Conversation* is just myself, Derek and the actors. Even his feature films don't follow the conventions of narrative feature filmmaking, in that there are no reverse shots, or any of the language of cinema. He basically shoots them as he would shoot a Super 8 film. At that time, if you wanted to make more ambitious work, you wouldn't be allowed to shoot it on Super 8. The larger the scale the production, the more constraints there were, and Derek balanced the two things, which is why we made *Last of England* and *The Garden*. That gave us the freedom to work in the way

(L–R): Journey to Avebury *(1973),* Andrew Logan Kisses the Glitterati *(1973),* In the Shadow of the Sun *(1972–1974).*

INTERVIEW

that other avant-garde filmmakers worked, but still have the product that the financiers needed to put in the cinema.

Where were these films showing?
He had a big studio so he'd show them there, and then he took them places like the ICA and the odd festival. With Super 8 he would actually take them to the festival. It's not as though they were distributed in any way. That's why we started working together. I was able to get funding to have some of the work transferred to 16mm, so it could be sent in the normal way to a festival and they could project it.

When were you blowing those up to 16?
Apart from the Co-op, I was programming for the Expanded Cinema section for the Forum in Berlin. I persuaded them to pay for a 16mm blow up of In the Shadow of the Sun, which premiered in Berlin in 1981. Then we went on to another three or four shorter films onto 16mm. It's quite expensive and it wasn't easy. Then video started to come in, so we just started transferring them to U-matic.

Was there a batch of LGBTQ festivals in the early eighties or do those start up later?
That was a bit later. When we were showing in Berlin, at what is now called The Panorama, [that] was in many ways the prototype gay and lesbian film festival. We had had a gay film festival at the Film Co-op a bit earlier, maybe 1980. But there wasn't a big circle. I made a film with Ron Peck called What Can I Do with a Male Nude? in '84. The BFI put a small amount of money into it, so therefore had the rights to distribute it, but they wouldn't put it in the catalog because it was too gay. They just had it listed as the Ron Peck film. It wasn't hostile to gay and lesbian work, but it was seen as very risqué. Interestingly within about two years of that, the BFI production board had practically turned into a gay and lesbian film

Derek Jarman from B2 Movie.

factory. Then the whole thing started with, was it Sundance, with queer film? It was just a marketing tool. I'm skeptical about the whole thing. Rosa Von Praunheim, Ron Peck, there were a lot of people making films that would be seen as queer films these days.

Can you talk about the challenges of restoring some of these films? Because of the way Derek projected things, are you having to make choices about how to transfer things?
I worked with the same guy that Derek and I worked with from the mid-eighties, Tom Russell. We're scanning at 2k, because we reckon that was enough for Super 8. We carefully transferred all the material, and then worked using Baselight, an advanced grading system. From that we kind of restored, kind of conserved some of the films. I show some people who were there at the time, and they'll say, "Well that's a bit fast," or "That's a bit slow," because you know the Bolex wasn't running at precisely three frames a second, or nine frames a second, whereas the Baselight does. So judging is hard. I mean we had to make a start, because the stuff that wasn't shot on Kodachrome 40 was fairly purple. If we hadn't had Baselight, that level of imaging software, we wouldn't have been able to restore it. If you just project the films on a Super 8 projector, it would have just been purple splotches. The Ektachrome from the early seventies is quite notorious, isn't it? And then he was using print stocks.

He was shooting with print stocks?
It looks like it. I made notes of all of the different edge numbers. But apart from Ektachrome and Kodachrome there were other stocks as well.

What was his black and white of choice?
I think Tri-X was his black and white of choice, although he used Plus-X as well. He occasionally used 4-X, It's a good picture Tri-X, isn't it? ■

INTERVIEW

Paul Sheptow

FOUNDED BY PAUL SHEPTOW in 1972, *Super-8 Filmaker* magazine was aimed at artists and independents eager to give Super 8 a go. The magazine was chock full of ideas for Super 8 filmmakers who wanted to take their craft seriously and get the most mileage out of small gauge equipment. Flipping through the pages of *Super-8 Filmaker* is an absolute treat. Sheptow brought on top flight designers; consequently, the magazine is a feast for the eyes, filled with great photographs and vibrant illustrations. Let's not forget that the early 1970s had a certifiably groovy aesthetic, which is on full display in the magazine's pages. The publication also serves as an archeological excavation of forgotten media; there are plenty of equipment roundups, demonstrating how many companies attempted to capitalize on the Super 8 phenomenon. You'll see ads and articles about almost every camera, projector, splicer and viewer to hit the market, as well as for oddities like helmet cams, underwater rigs, belt pods, the Cut-A-Rut, which removed unwanted sound stripe from your film, and my favorite, the Hedler Ventilux Lamp Helmet, which basically sets 1,200 watts worth of high-intensity lights on your head!

The magazine also featured festival round ups, articles about filmmakers, and tips on how to shoot, edit, and animate. Prominent Super 8 filmmakers like Lenny Lipton and Dennis Duggan were frequent contributors. The Fall 1973 issue features my all-time favorite article about filmmaking, "It's a Super 8 Baby!" Indispensable advice is offered in this one, including: "The prenatal classes will often show movies of births. See as many as you can. If you can't stomach these films, you should question your ability to shoot the real thing."

Super-8 Filmaker was also aspirational. There were tons of articles about using Super 8 as an educational tool in the classroom, as well as countless articles about how to create your own production company utilizing Super 8 gear. At its height, *Super-8 Filmaker* had print runs of roughly 50,000. It started off in New York, publishing five issues the first year. In 1976, the magazine relocated to San Francisco, and was up to eight issues per year before it folded in 1981. Over its lifespan, *Super-8 Filmaker* published sixty issues, at which point Sheptow changed the name to *Moving Image*, incorporating video coverage into the mix. *Moving Image* lasted just one year, folding permanently in 1982.

What made you decide to start a magazine geared toward Super 8 filmmaking?
I was about thirty and my fiancée and I went over to the Middle East to travel. She was a still photographer and I wanted to also record our journey. Since she was taking stills, I thought I'd make Super 8 movies. Her uncle used to be an executive at Ehrenreich Photo Optical, a distributor for Nikon. I went and asked him if I could get a good deal on a Super 8 movie camera. So I went abroad and shot Super 8 movies as an amateur. I came back to the States and I was looking for information on editing Super 8, and was realizing there wasn't a lot out about Super 8. I had just graduated from Wharton School in marketing, so I have a business background. My mind went to, "My God. Here's a need. I have a need, maybe others have a need." Mike Matzkin and Tony Galluzzo were two writers for *Modern Photography* magazine. They would have articles about Super 8 in what was then the largest photo magazine in the world. I called Mike up. I said, "There's a need for a Super 8 magazine." I started putting it together and he agreed to write for me. John Lidstone had a book out about Super 8. I wrote to him and he agreed he would write. Everybody I wrote to who was in the Super 8 field all jumped at the chance to write for a Super 8 magazine. One of the big things was I happened upon a guy named Kit Hinrichs. Kit had an art studio and agreed to be my art director on a part time basis. Graphically it's won a lot of awards and I'm quite proud of the way it looked. Now he's one of the partners in Pentagram, one of the best design companies in the world. We had some very famous art directors like Chris Blum who was in charge of the Levi Strauss account for some advertising agency. These guys work for big companies, so he can go out and get Levi's to pay thousands of dollars to have a piece of art done. Well we can't afford that. But he could ask that same artist to do us a piece of art for $200. That's how we had some of the best illustrators in the entire country.

The magazine comes out in 1972. Is there a nice response right off the bat?
Yes. I pre sold it. This is my first magazine, so I thought I had to get the advertisers. I

flew out to California just to get Beaulieu. One advertiser—a totally inefficient trip. But once you land one advertiser, you have them for life. Kodak traditionally bought every back cover of every photo magazine and I thought that they would buy mine. On my dummy that I took to advertisers, I put Kodak on the back cover. Everyone would say, "Did Kodak buy the back cover?" I'd say, "Not yet. I'm working on it." I went to Kodak and they put me through the ringer. I had about five different meetings with the company and the fellow in charge of the Super 8 at the time told me, "Don't be sad if you don't get advertising from Kodak because Kodak doesn't normally go into new issues. They wait until you prove yourself." It's under that dark cloud that I was there selling. I presented and it was a good meeting but as I'm leaving [they] turn to me and say, "Mr. Sheptow, you know Kodak spends a lot of money advertising. You really think that us running ads in your magazine is going to make any dent in our profitability or in our sales?" I basically told him, "Maybe for your Kodachrome II," the most commonly used Super 8 film, "Maybe we won't." But they had a new Ektachrome 160 high-speed film, and I said "I could probably guarantee you that we will help you sell Ektachrome 160 because you're introducing that film. If anybody's going to buy it, it's going to be our readers." When I got home after that trip I got a call from their ad agency and they said they would like the opportunity to buy all the covers for the next year and run other ads inside the magazine. *Modern Photography* and *Popular Photography* would rent me their photo lists, so I did direct mails. And *American Cinematographer*, I did real well with their lists.

How many magazines are you selling in the seventies?
We were printing 50,000. We put a lot on the newsstand and we had a lot of subscribers. We were at that time the world's largest-selling filmmaking magazine. There was only *Movie Maker* in London and *American Cinematographer*, which was a professional magazine. We had a booth at Photokina. I traveled around the world promoting Super 8. I was in Africa where I was interviewed on the radio about a Super 8 Festival in Africa. I was at the Caracas Super 8 Film Festival. I helped spread the word about Super 8.

The magazine is definitely serving as a pied piper for the medium. Who was the Super 8 community in your opinion?
The serious amateur, independents, and professionals. That was our audience. There was an independent 16mm magazine, but there was a group of 16mm people who were dabbling in Super 8 and then would blow it up to 16mm. That's when Super8 Sound started. There was a place in San Francisco which started blowing up Super 8 to 16mm. Dennis Duggan in San Francisco was very involved in doing that Super 8 to 16mm stuff.

Super 8 was seen as this international phenomenon. Did you sense there were different styles of filmmaking coming from different countries?
I didn't. But I think they were all going through the same growing pains of Super 8. All the Super 8 aficionados were waiting with bated breath for the next creative invention that would move Super 8 to another level and they looked toward our magazine to bring them that news.

There were a lot of articles about monetizing Super 8, turning it into a professional business, using it for commercial and industrial uses. Did that ever take off?
No it didn't. They were just working on the technical aspects of being able to blow it up, being able to have good sound and editing. All that was coming in when videotape came in. If you can imagine a graph and you see Super 8 going up and then videotape starting and video went straight up. That made Super 8 go down, and it was before they really made the technical ability good enough to use Super 8 professionally.

"All the Super 8 aficionados were waiting with bated breath for the next creative invention that would move Super 8 to another level and they looked toward our magazine to bring them that news."

INTERVIEW

There seemed to be a lot of articles about getting Super 8 into the educational market and into the high schools and elementary schools. Were you sensing that was happening as well?
Definitely. We we asked Yvonne Andersen, who was an animator in schools and teaching in schools, how to do animation. Look, before videotape what kind of moving image mediums did you have? You had 35mm, 16mm, 8mm, and then Super 8. Regular 8 didn't have the technical capabilities, and 16mm was getting pricey and professional. Super 8 had a niche in schools, in independent and in semi-professional moving into professional if that gear could have gotten up to that quality. We were there. It was happening and then videotape came in and just stopped it in its tracks. We started to see some Super 8 films blown up being in festivals. If that would have continued, you would have had more acceptance of it, more creativity, more highly technical stuff [with] better quality. Every place you can imagine it—education, industrial, professional, weddings. It was a way for people to monetize filmmaking and not have to spend a lot of money. Especially when Kodak came out with Ektachrome 160, and then double system, it started getting professional. But again, all that was stopped. Super 8 was on its way to more respectability and it was snuffed out by video. I then turned *Super-8 Filmaker* into a magazine called *Moving Image*. I sent letters to our readers saying this is the direction where a lot of our users are starting to get into videotape. We covered both Super 8 and video. Of course we got some people upset. You know we felt we had to chase the market or we basically would have had a magazine about buggy whips. But the interesting thing is the first revolution of videotape was not people who made videotapes. It was people who wanted to watch videotapes at home. Betamax and VHS.

We actually made the switch, but it was too early for video making. We were ahead of that curve and watching our other side die. ∎

Jonathan Tyman

AFTER GRADUATING St. John's College in Santa Fe, New Mexico in 1977, Jonathan Tyman moved to Los Angeles to explore his interest in film. He landed a job at a special effects company, but his developing interest in experimental film led him away from Hollywood. In his mind, Ann Arbor was the place to be. This Midwestern college town is home to the Ann Arbor Film Festival, the country's oldest experimental film festival, founded in 1963. The Ann Arbor Film Fest, however, only accepted 16mm films. In its shadow, lived the Ann Arbor 8mm Film Festival, which was open to smaller gauges. Founded in 1971, it lays claim to being the first festival of its kind in North America. The Ann Arbor 8mm Film Festival took place on the University of Michigan campus. It screened films by Rocky Schenk (see Rocky Shcenk interview, page 160), Marc Huestis (see Marc Huestis essay, page 126), and Oscar winners Alex Gibney (*Taxi to the Darkside*) and John Nelson (visual effects master for *Blade Runner 2049*, *Gladiator*). Of course, at the time no one could be certain if any festival participant would ever make another film, let alone have a storied career. Todd Haynes was recognized by the fest in 1979 for his short *The Suicide* (available on the Criterion edition of *Safe*), which took home an honor and a cash prize of a hundred dollars!

The presence of the prestigious Ann Arbor Film Festival suggested to Tyman that the University of Michigan was a hotbed of alternative film production. Once on the ground in Ann Arbor in 1980, he found this was far from the truth. Ann Arbor was an amazing place to watch films, but almost nobody was engaged in the type of filmmaking on display in both the 16mm and 8mm festivals.

While at Michigan, Tyman enrolled in a newly-formed graduate program in Telecommunications. That program also turned out to be less glorious than anticipated, as the students' access to equipment was limited to television cameras from the 1950s. In 1981, Tyman took over film instructor duties in the university's tiny undergraduate program, where he focused on moving the department from a film appreciation program to one that offered production courses. Though progress was slow, under Tyman's watch, the program grew, eventually becoming a well-funded undergraduate film program by the late 1980s. Tyman also became a regular at both of the city's independent film festivals. He served on the screening committee for the 8mm festival, and took over Festival Director duties in 1985. He currently serves as Board President for the Ann Arbor Film Festival. On a personal note, Tyman was the first person to put a Super 8 camera in my hands, having taken his first-level film production course at University of Michigan in 1985. He was the instructor whose softball game I visited in order to borrow a sound projector, as recounted in Chapter 2.

INTERVIEW

Where were you prior to Ann Arbor?
I was in Los Angeles. The 16 [film] fest might have been the major reason I came to Ann Arbor in 1980. I was drawn by the illusion that the center of the avant-garde film world was Ann Arbor. It turned out Ann Arbor was a great place to watch movies, but there was no momentum toward production. I was working in 16mm, but I was hanging out with the people who were running the 8mm Film Festival and I got sucked in. There had been a great scene in Ann Arbor prior. Are you familiar with The Once? Have you heard the term the "Happenings"? Well this is where they started, and it was The Once, a.k.a. The Once Group. They didn't like the group part because they wanted to emphasize that each thing they did was unique and each event was a constellation of characters that would never assemble again in that particular form.

That's more in the art-performance-music world?
Yeah. Performance and avant-garde. It was a Noah's Ark of people that had come out of the various arts including architecture, dance, painting, sculpture, and film. George Manupelli was a part of that group. He's the one who started the Ann Arbor 16mm Film Festival in 1963. '62 to '69 was a real culture-breaking time in Ann Arbor. But by the time I arrived in 1980, there was Cinema Guild and all these places to watch movies, and you could see avant-garde film, but there was almost nobody making anything of that sort. Really, there was a worshipping of Hollywood, which is where I had just come from.

Were you making your own films at the time?
I was making in 16. I was playing in 8, but I didn't take that seriously–unlike much of the world, as in Europe, South America, and Africa.

Where you think Super 8 was being treated more seriously?
It was the 16mm of those places because it was much more affordable and controllable in terms of being able to send your footage to be processed and so forth. I found a notebook from the 15th Ann Arbor 8mm Film Festival [from 1985]. It's the names of the film, the name of the filmmaker, where they were from, and then our few notes, and the number score that we gave them. There were over 200 films. Over 10 percent of them are from outside the US twenty-four of them. They're from places like Hungary. Three of them from Algeria. Three from Finland. Four from Germany. I remember those were some of the best films because they were actually interesting. They may have been just glimpses, but they were glimpses of something that we weren't familiar with. I would say that Americans played with it more. The bar was lower so they could do anything they wanted. But overseas, the things that came in tended to be either narrative or documentary with a heavier leaning toward documentary.

Are you suggesting they were complete films as opposed to experiments in style?
Yes.

At the time would people be sending VHS copies or prints or the originals?
Most of it was original. With tape splices. It was very analog. We were running it through projectors that we knew intimately. There was a big [Elmo] and a little one. The big one was used for the screenings. The rule was every film was run through start to finish in the room where the screening committee was present. That didn't mean that we paid attention to every film. Some of them were really bad. There would be seven to eleven people on the screening committee. [It] was ad hoc. Whoever was in the room was on the committee. And then somebody was taking notes. [Editor's note: Tyman reads one such note about a submission.] "Young man dreams of a beautiful woman in white. She eventually materializes and awakens him with a kiss." That film has been made by every student. [Reads another] "Man spends his last hour at a typewriter with intentions of rectifying matters. Flashbacks."

Where were the screenings held?
They were held on campus and they filled up. I know that once it was in the MLB4 [Modern Language Building] which holds 400 people. To be honest it was more fun and a bigger party than the 16 Fest. One distinguishing factor was it was international. We hosted twelve, maybe more, festival [directors] from overseas. Venezuela sent a couple of people and they would hand carry films from their festivals. My first overseas travel in 1981 was to Lisbon because the festival there was run by two guys who were airline

> **"I liked the people that surrounded the 8 fest more than the people who were involved in the 16. There was less pretension. There was more fun."**

INTERVIEW

stewards and they got free flights. The city of Lisbon gave them all kinds of money to run it and it was a gratis trip to Lisbon. I did the same thing that they would do, bringing a reel from here. Two of us went over there and traveled around and shot film and partied for a week.

What was that festival like? How did the films differ from what you were getting in Ann Arbor?
In Lisbon? Well most of them were in languages I did not know.

That's right. No subtitles in Super 8.
No, they weren't subtitling, so I was in the dark a lot. But in another way, not. The language of film was pretty universal. I would say there was fiction and documentary. Nothing of the experimental leaning that I would have been excited by. The people doing that were using 16 more.

I should mention that I noticed that the largest single venue from which the submissions came that year were from Southern California. I remember one of those films had a ton of special effects in it. They sent a print. Some of them were just absolutely stunning. There were also the GI-Joe and Barbie genre that plagued Super 8 at the time.

People reenacting things with GI-Joes and Barbies?
Yes. For some reason they really liked to do that. They would always include a little pixilated sequence. Often, they just took them by the hand and walked them around. They were usually dirty, but some of them were on the opposite end of the spectrum and were just staggering. I remember there was a Yogi in meditation, and they did a matte of his forehead and had a triangle out of which grew a whole imaginative world. It was remarkable. They were doing mattes and keys in Super 8, so they probably worked at a lab. I can tell you about the labs. You could get your footage processed in three different labs here in Detroit. There was in general less care taken with Super 8 than with 16. A certain number of [rolls] would come back where the person, instead of breaking the cartridge, had pulled the footage through the felt, and static electricity built up and there was lightning throughout the film. You're just going, "Well, that wasn't intended."

How did you feel about what you were seeing at the 16 versus the 8mm Festival? Was there a palpable difference in the type of work?
The 8 fest had more traditional stuff. But it was on the cheap. It was more personal. It was, in a way, more interesting. I liked the people that surrounded the 8 fest more than the people who were involved in the 16. There was less pretension. There was more fun. The 16 fest was full of artists who were breaking the culture. They were very self-indulgent. They were also making pornographic films and showing them in public. The 16 Fest was getting sued for it with the consequences actually changing legal precedents. So there was a bigness to the 16 fest, as compared to the 8 fest, culturally and artistically, as well as just in terms of the size of the image. Some of the 8mm did have actors and costumes and stuff, but very little of that. If you're going to go to all that trouble and expense, it usually wound up on 16 or even 35.

Can you talk a little bit about how you ended up teaching at Ann Arbor and what the program was like?
There was Film/Video 101 and 102. One was film and one was video. They were both art appreciation courses. There had been no making in either of them. Martha Schmidt was teaching video. I was teaching film. Both Martha and I were leaning toward the making. I was not interested in having it be another movie watching course anymore when I took over Film 101.

When did you take that over?
Probably 1980 or '81. Nobody was making anything and it may be that all of the cameras you had access to and projectors and editing equipment I bought personally. I bought probably fifty cameras during those five years because they would come back broken. Some of them were really good—a Beaulieu…

I can tell you I didn't use that at Michigan.
Right, but I did finally get one. It was like a hundred bucks. For the most part, what I found was that people had better results with fixed focal length lens, automatic Bell & Howell point and shoots. And the editing was just those little splicers. I think we were probably for the most part using wild sound on cassette. Does that sound familiar?

The first film I shot was silent so that could be right. And then the rest that I shot were single system sound, but I borrowed a camera from a friend. I remember that the only projectors we had in class were silent. If you made sound films, you couldn't really watch them. I remember at some point meeting up at softball game to borrow a projector you had in your trunk.
Danny, you have the whole tone of the festival. The projector would be in my trunk. That sounds right. And for all I know I would have just bought it. I would drive to Chicago and to Detroit to the used camera stores and buy up all their Super 8 stuff.

Can you talk a little bit about the Venezuelan festival?
That's where I met Lenny [Lipton]. He was sick as a dog. He would not get out of a hammock except to puke. It was a real bummer. He was sort of the god of Super

158

INTERVIEW

15th Ann Arbor Eight Millimeter Film Festival
February 6-10 1985

8 from my perspective. It was a big festival. Everybody who was at the festival was a part of the festival. It wasn't like we treat it here where you have a festival and you invite the community. There were sixty, eighty of us, all people connected to the festival and it was state sponsored. We were put up in homes and bussed around for entertainment for a week. It was a week-long party and we'd watch movies at night in the dark and stay up really late.

Was Carlos Castillo the head of the festival?
That's him. He also was a performance artist and I helped him do a performance up here that I still remember because I had to go get a bath tub and seven cans of shaving cream and mount a projector shooting down on the bath tub with a coating of shaving cream on it and he was underneath the shaving cream naked. And the film was of him taking a shower that was projected on the shaving cream. Near the end of the film he bursts out of the bath tub. It was pretty funny.

At the University of San Francisco I teach one 16 class and then a class where we shoot Super 8.
I have noticed a resurgence of interest in film, but I got over it a couple of decades ago. It was a real hard transition going from film to video.

It's tricky because you have to learn an entire new way of making things.
And a whole new way of being. That's actually what interests me the most in terms of what I reflect on. Does the medium and the aesthetic that goes with it have its own character? I think it does.

And I think with Super 8 or 16, through the seventies and eighties and nineties, the cost was so prohibitive that you would shoot black and white in 16 because it was cheaper. You would shoot without sound and create some kind of soundtrack later because it was cheaper than shooting double system. Or you'd work in Super 8, and that meant everything was straight cuts, and no soundtrack was going to be more complex than two tracks. The medium is definitely affecting the aesthetics.
The difficulty of it was so tremendous and burdensome. You were just dealing yourself a very hard hand to play if you wanted to have that level of sophistication. The more interesting thing to me is, if you increase your limitations, what does that do for you? My mentor was named Rudolf Arnheim. He was the defender of the silent movie, thinking that parsimony increased, didn't decrease, expressive power. There is absolutely something to that. The limitations that you struggled with, that you embraced, actually do, if you can tune-in to that aesthetic, have that intensity of expressiveness. ∎

INTERVIEW

Rocky Schenk

FOR PEOPLE OF A CERTAIN age, Rocky Schenk is best known for his 1980 Barnes & Barnes "Fish Heads" (1980) music video. It's a great song and a cult classic that Schenk codirected with his friend and collaborator Bill Paxton. The Super 8 films Schenk made before that in the 1970s were absolutely exquisite, simply some of the most beautiful Super 8 films you'll ever see. Not only do they look good, but they were big productions. *The Egyptian Princess* (1977), was shot in a Hollywood mansion, and features a huge cast, with great attention paid to art design, wardrobe, and make up. There's even a crane shot! The rigor of the editing in *The Egyptian Princess* and its predecessor *Dream Sequence* (1975) firmly dispels the notion that sophisticated work couldn't be made in Super 8. These aren't films made by someone goofing around. It's clear, Schenk was an artist taking himself and his craft seriously. In the 1970s Super 8 landscape, Schenk made a name for himself. His films netted several reviews in *Super-8 Filmaker* and his name was all over the program brochures of the the Ann Arbor 8mm Film Festival. Schenk went on to a successful career as both a photographer and filmmaker, shooting music videos for everybody from Devo to The Cramps to Nick Cave & PJ Harvey to Ace of Base, Gloria Estefan, Van Halen, Alison Krauss & Robert Plant, and Adele.

When did you start making movies?
I was born in Austin, Texas and raised on a ranch in the hill country about eight miles from a little town called Dripping Springs. I spent a lot of time watching classic films on TV. My parents were big movie fans, and would take my sister and I into Austin regularly to see the latest films. We saw all genres—horror, animation, comedies, dramas, musicals, epics. I started writing and directing plays and variety shows when I was in elementary school, and my class made a little film of one project around the fifth grade. I made my first "official" film when I was about fourteen or fifteen—a short musical where I attempted to duplicate the style of musicals from the 1930s, foreshadowing my future directing music videos. The film was shot in black and white, and had a somewhat elaborate opening credit sequence where each letter on each title card "blew" into place—achieved by filming the sequence upside down and blowing the letters away, then editing the film right side up and reversed.

Were the first films you shot on Super 8?
Yes, Super 8—it was my dad's camera and equipment. My dad loved making home movies of the family, and occasionally would let me use it to shoot projects of my own. I used to build miniature structures out of cardboard, and then burn them up, capturing the destruction on film. Once in a while, he would create short films starring me and my sister, filming with his Yashica Super 8 Camera—I forget the model but it was similar to the SU-40E model. This was the camera I used, and that I would eventually take to college and then to Los Angeles.

Was Super 8 a conscious choice or was it simply what was available to you?
I had my dad's camera and equipment available, so that's what I used. I studied and read lots of books about movies. Living with classic films for so many years, I instinctively knew how to direct, light, shoot, and edit. My Dad also bought me a Yashica Twin Lens Reflex still camera, so I began learning still photography and filmmaking simultaneously. I began shooting stills on the sets of my films, which would later lead to many years of photography jobs.

Dream Sequence **is pretty advanced in terms of its choreography, art direction, and editing. At this point, were you actively thinking of yourself as a filmmaker?**
Yes. I knew after my very first film—the retro musical short—that I wanted to pursue filmmaking. I was also an artist and a painter during my teenage years, and started selling my work when I was around thirteen. But I was addicted to movies and writing little stories, so I kind of knew where it was all heading. I went to college at North Texas University in Denton, TX and majored in Art—to make my parents happy. I knew they would not be as supportive of me majoring in film—and I suppose they thought I had a better

The Egyptian Princess *(1977) with Bill Paxton.*

INTERVIEW

chance making a living utilizing my artistic abilities—advertising art, drawing, etc. Also, I remember asking them if I could move to Hollywood when I was a teenager, and they gave me a firm "No!"

At college I began writing and shooting my second film, *Dream Sequence*. During this time, I met my first filmmaker friend, who was moving to Hollywood to hook up with his friend Bill Paxton, who I didn't meet until I moved to Los Angeles. My friend wrote letters describing his adventures in Hollywood with Bill and encouraged me to quit college and move there, which I did about a year later. It was the best advice I've ever had! I worked on *Dream Sequence* for many months. The actress in the film was a fellow art student, and my sometime "crew" consisted of a couple of my college roommates who pushed me in a wheelchair for my "dolly" shots. The extras and dancers were from the dance department. I knew exactly what I wanted to accomplish with this film when I began writing it. Since about age five, I had been having a series of recurring nightmares, and I wanted to visualize on film my experiences. I don't want to get into it too much here, but I would wake up crying and screaming after these dreams and my parents would always be by my side, attempting to calm me down. When I moved to college, the nightmares continued. I remember scaring my roommates by screaming at the top of my lungs in the middle of the night. I finished filming at North Texas, quit college, and moved to Los Angeles. I worked for months editing the film on a primitive little editing system with a small viewing screen, and began listening to a variety of classical and experimental music for the soundtrack. It was "learn as you go," because I had not taken any film classes while I was in college, and I was depending on books and my few filmmaker friends to provide me with technical information. I was still using my Dad's Yashica Super 8 Camera for all of these early films, but I would occasionally rent another camera that allowed me to do "in-camera" fades and dissolves. I used tape splices, because I thought they looked smoother than the "glued" splices. Film stocks for black and white were either 4-X or Tri-X. I remember being very concerned with keeping the film in good condition and using gloves while editing. I knew this was my original "master" for making future copies.

How did you do the superimpositions for the end title sequence?
For *Dream Sequence*, I stuck white letters on pieces of glass and illuminated them with a side light, then projected edited sequences from the film at 12 frames per second on to the letters and a screen behind the glass, then filmed it at 24 frames per second with the camera, creating the illusion of a slow-motion superimposition.

How did you record the soundtrack?
That was a complicated procedure, because I had to create the soundtrack all in one session without mixing equipment. I had multiple cassette players and a turntable, so I had everything all cued up and kept changing records and cassettes while I recorded it with a separate Sony cassette recorder. It took forever to get a smooth pass! I eventually had the film copied and added the prerecorded track to the magnetic stripe on the copy.

After finishing *Dream Sequence*, I started my third film in Los Angeles, another psychological drama called *Killer Chihuahuas*, also shot in Super 8. In a nutshell, the story was about a middle-aged, overweight, dissatisfied single mother with two kids who lives in a grim, lower income home. She escapes her depressing reality through her rich fantasy life, and her dreams are visualized

Rocky Schenk, Bill Paxton and a pile of Sun-Guns.

throughout the film. *Killer Chihuahuas*, about thirty minutes long, played at Filmex (the Los Angeles Film Exposition) and got some nice reviews—including a write up in either *Variety* or *Hollywood Reporter*. I was thrilled! It also played in a few other film festivals including the Ann Arbor 8mm Festival, where together they won the top prizes. I cherished the letter I received from the festival, and still have it on my wall in my office. "Highlighting the festival were two films by Rocky Schenck of Los Angeles… mood pieces of exceptional composition and form." *Dream Sequence* also won first prize at the "Amateur 8 Movie Contest."

Egyptian Princess has such high production values. You are clearly not looking at Super 8 as an amateur medium.
I had a vision for this story and a fascination with this strange new city where I now lived—Los Angeles. I was from a ranch

outside of a small town in Texas, so Los Angeles was like one big "set" for me to film and explore. I met Bill Paxton, who had moved to Los Angeles the year before and was already working doing various jobs on different low budget films. We hit it off instantly and became close friends, getting a small apartment together in Echo Park. Bill had dreams of being a movie star as well as a filmmaker, and I dreamed of being a writer-director. We both had multiple jobs during this time. I designed movie posters for low budget films and worked as a "gofer" at a film production company. I also drew and designed the newspaper ads for Pussycat Theaters and did my very first professional photo shoot with a "star" of one of their films. Bill worked on the film *Carrie*, finding locations for production designer Jack Fisk. I remember visiting the set of *Carrie* with Bill, watching them filming the sequence where Carrie is locked in the closet with her mother, and the scene where she walks through the gymnasium toward the stage, on her way to being humiliated with a bucket of blood. I can remember shuddering watching them film that sequence. It seemed familiar to a humiliating incident that I had experienced in my youth in the gymnasium of Dripping Springs School.

In between working at our various jobs, we filmed *Egyptian Princess*. I was pretty much a one-man-crew. I wrote and directed the film, drew the storyboards, and did all the cinematography and lighting—utilizing seven 650-watt "sun-guns." A friend made the bride and bridesmaids costumes, and I made the Nefertiti hat. The plot of the film concerns a disoriented woman searching for her cheating lover in exotic locations around Los Angeles. Designed as homage to classic silent films from the 1920s, the delusional woman imagines the man having a series of rendezvous with beautiful women in an abandoned mansion, a dinner party, an art deco chandelier, and several other environments. Bill coproduced the film and I financed it. We utilized all our friends as actors and extras, and another friend helped "build" the soundtrack with me. I had various friends push me around in a wheelchair for the dolly shots, or drive me around for exterior tracking shots while I sat on the hoods of their cars. I executed my first "crane shot" thanks to the production company where I was working at the time, who volunteered their crew and their soundstage.

I didn't pay for a single location. Most places granted permission after learning that this was not a commercial film project. There were a couple of locations I wanted, and we broke into or snuck into a few places without permission. I wanted to shoot inside the Pantages Theater in Hollywood for one sequence. Bill and I were friends with a female usher who worked there, and the usher would sneak us into the last screening of the night, and we would hide in the back row of the balcony while everyone left the theater. We'd parked nearby, so when the coast was clear we would bring our lights and equipment into the theater and shoot until sunrise.

While editing, I came to the realization that there were recurring themes of betrayal, humiliation, madness, and duality in each of these three films—themes that were based on an unfortunate incident that I had experienced in high school and had completely blocked from my memory. It's too much to go into here, but this "repressed memory" motivated me to create these somewhat psychological dark films, and provided me with a profound "drive" to get these stories out of my system. Filmmaking as therapy, I would call it.

What possibilities did Super 8 offer you?
It allowed me to make films by myself and without a big crew, and to visualize the stories that were in my head. At that time, I saw absolutely no limitations or weaknesses with Super 8. I could make films and that's all I cared about. And I couldn't afford 16mm at the time—way too expensive. Working with Super 8…well, I probably wouldn't have become a filmmaker without it!

How did you move up through the filmmaking ranks?
Non-stop work and word of mouth. Folks would see my work and either recommend me to other folks or hire me themselves. I was lucky to meet some extraordinary people who believed in my talent and who championed and encouraged me throughout the years. Without them, my career wouldn't have happened. I have to credit Bill Paxton for changing the course of my career, and perhaps my life. He asked me to codirect our very first music video together, "Fish Heads," for Barnes & Barnes. I shot and directed all of the opening sequences before the animation sequence and the song begins—about the first two minutes—which were all shot in Super 8. I shot and codirected the color "fish head" party sequence, also Super 8, and painted lipstick on the fish heads… and added their false eyelashes. I also shot and directed the fish head garden sequence, and the majority of the scenes featuring Barnes & Barnes dressed in trash bags and goggles—mostly shot on Super 8. Somewhere along the way, we transitioned to shooting in 16mm, with various folks helping with the cinematography and Bill directing sequences on his own when I was unavailable. Bill took "Fish Heads" to New York and loitered outside the offices at NBC until they took a look at the finished clip. When the video premiered on *Saturday Night Live* with Bill and I sharing codirector credits, it really opened my eyes to all sorts of new possibilities. This was pre MTV, where the video would soon be seen regularly. Bill's acting career started to take

INTERVIEW

off, but we continued working together on additional videos for Barnes & Barnes. I took over directing, filming, lighting, and editing the videos, all in 16mm by this time, and Bill would star in them.

I was seduced by the creative and artistic freedom I encountered in the world of music videos, and I'm wildly thankful for the diverse group of artists, cinematographers, editors, actors, gaffers, grips, art directors, make-up artists, hair stylists, costumers, choreographers, and extras I've been able to work with over the past several decades. This unique artistic medium has challenged me to use my imagination to its fullest, and has made it possible for me to write and direct over 150 "short films."

Can you share any classic Super 8 war stories?
For *Egyptian Princess*, I filmed one sequence requiring many extras that took about fourteen hours at a location that was rather difficult to acquire. When I got the film back, there was a continuous large scratch down the middle of the film ruining an entire day's work and requiring a total reshoot. The lab gave me credit and replaced the film, but the location we had used would not allow us back into the building to reshoot. Undeterred, we rescheduled the talent and extras and showed up very early at the location and broke into the building, climbing through some windows. We refilmed the entire sequence without the owners of the building knowing we were ever there.

Another time, Bill and I were in downtown Los Angeles scouting locations at night when two guys came up to us and started pushing us around and asking for money. They knocked my glasses off, pushed me to the sidewalk, and roughed up Bill. They got some money, but they didn't ask to look in the paper bag I was carrying, which held my cherished Super 8 camera! ∎

James Nares

THE MYTHOLOGY OF NEW YORK in the early 1970s is of a crumbling city on the verge of bankruptcy. Into that milieu fell a band of musicians, filmmakers, painters, and performance artists whose edgy work reflected their troubled times. James Nares was an important part of that scene. He was the original guitarist of The Contortions, one of the era's seminal No Wave bands. In addition to making that noise, Nares picked up a Super 8 camera and shot an impressive number of films between 1975-1980, becoming part of a filmmaking scene that included Eric Mitchell, Becky Johnston, Beth and Scott B, Vivienne Dick, Amos Poe, and Jim Jarmusch. Immediate, rough-hewn, and petulant, No Wave filmmaking embraced Super 8 and shared the young, loud, and snotty vibe forged by their punk rock brethren. Nares' film work was quite varied. He studied bodies and bodies in motion (*Block, Arm & Hammer, Steel Rod*), made documentaries (*No Japs at My Funeral*), and created experimental work (*Waiting for the Wind*). His most ambitious piece, a feature-length narrative, *Rome '78* stars scene stalwarts John Lurie, Lydia Lunch, Pat Place, and Lance Loud. Nares was born in London, grew up in Sussex, and moved to New York in 1974. He continues to work in video today and is also known for abstract paintings that maintain the vigor and energy of his Super 8 work.

You went to art school in England correct?
I went to Chelsea in London, but I didn't stay long. I came to New York because I was interested in American artists and the New York scene. Nobody in the art schools I went to knew about the things that I was interested in. I lucked out when I did come here because I stayed with a friend who had just rented a loft in Tribeca, in the days when it was a ghost town and when it was the center of what was happening in New York.

When you arrived in New York did you enroll in school?
I enrolled in The School of Visual Arts, and I was there for two semesters. I'd used it as a way to get a visa and as something to do. I had no idea how the American education system worked. I'd come from a system where you showed up and you were told what to do. "Go here, go there, sit down, learn this." It took me a while to figure out that you had to sign up for classes. You get to choose what you want to do—seems kind of novel to me. I didn't stay very long because I just figured, with a certain measure of arrogance, that I knew what I was doing.

What were you studying at SVA?
I did sculpture with Richard Van Buren and Hannah Wilke. I studied video with

Hermine Freed. It was sculpture and video because they were the only classes that were left. I did a number of things using Portapak. I was making videos with that lovely, enormous heavy equipment. I shot a movie, Eric Mitchell's *Bikers*, which was a nod to Kenneth Anger, and we shot the whole thing in Olivier Mosset's loft on Mercer or Crosby Street. We were into shooting movies in one day, one night. I remember there were about twenty people participating in the film and I was tasked with shooting it. It was a slightly debauched evening, and we shot the whole film in about three hours, and I realized that I hadn't tightened the reels. If they weren't just so tight, the thing didn't engage and the tape didn't run. So the tape hadn't even started running. In the great tradition of that style of movie making, we shot the movie again. That was a bad moment for me.

What was it about film as a mode of expression that seemed right to you?
It was Super 8. I used Super 8 when I was in England. I came back to it, but with slightly better cameras, when I was in New York. I loved Super 8 because it was so light, and I loved the three-minute units, or two-and-a-half-minute unit of the film cartridge. That cartridge was so sweet.

What was the first camera you came across in New York? Were you borrowing, renting?
I begged, borrowed, I won't say stole, but close. There was this guy on Houston Street, he furnished Super 8 cameras to a number of downtown New York filmmakers. They "came off the back of a truck" somewhere. But the camera I first remember using was a Beaulieu with an Angenieux lens. It was a really nice camera. I had it adapted to put out a sync pulse, so that I could shoot double system sound. I've been restoring some of the films that I shot with that camera, and they're unbelievably sharp at a 4K scan. I just restored one of my longer films from 1977 and we showed it at MoMA as part of their Club 57 show and it looks amazing. And the sound was shot with a different camera, the single system Nizo.

Tell me about working with double system.
Double system wasn't something I did much, because the single system cameras were coming out around then. Double system was clunky, and then you had to go through all the business of syncing it up. It seemed much easier to use single system. Then, with a lot of the projectors, you could record on the balance stripe, which opened up another whole world. I just I migrated to single system soon after getting this expensive conversion done on my Beaulieu.

Because you were in New York were you seeing a lot of the Warhol, Morrissey films or the Tony Conrad, Michael Snow universe? Did you see what you were doing as a complete break from that kind of work?
I was certainly aware of the Michael Snows of this world and Warhol too. I'd seen the Warhol films, the Paul Morrissey films and loved them. I would probably have identified more with the Michael Snows in the films that I was making with that minimalist aesthetic. But as Amy Taubin pointed out in the first review of my work, it was like I had that minimalist aesthetic but I added something personal. I forget how she put it, but a kind of angst. I married those two things together for some more expressionist kind of minimalism. I started off making things as a natural extension of the way I was thinking and as my contribution to what was going on. Later on I positioned myself more in opposition, as a lot of us did, around the time that I met Eric Mitchell and the punk rock scene. My first real attachment to that was I started playing music and I joined a band. It was that time when, as John Lurie says in Celine Danhier's movie *Blank City,* "Anybody who was doing what they knew how to do was very suspect." There was a great exuberance of doing different things and nobody had careers. It was a pure expression. It was also a pure antagonism because we were consciously in opposition to what we thought we needed to oppose. But it wasn't a negative opposition. We thought we were opposing for a greater good. It was like a cleaning of house.

How was the music scene influencing the films and vice versa?
The great thing about the music scene was its simplicity—the three-chord rock. We were making three-chord movies. The idea that if you want to make some music, just pick up a guitar and I'll play drums if you play bass, which is the way we approached the music. We would rent a rehearsal studio, go in there with a cassette recorder, put it in the middle of the room, and just start playing. We approached filmmaking with the same attitude. You just pick up a camera and do it.

"The great thing about the music scene was its simplicity—the three-chord rock. We were making three-chord movies."

INTERVIEW

How did Super 8 help this notion of three-chord movies?
How cheap it was. It was a lot less than any other way of making color, sound film. In the very early films I would shoot at five to seven frames a second and then project at the same speed so it had a kind of moonwalking. You could almost see every frame as it went by. You could extend a three-minute roll to like twelve minutes. So it was a financial thing. It was an ease of operation thing. I mean you could just hold the thing like a pistol and throw it in the air if you wanted. Those cameras were pretty simple and indestructible and portable.

You were shooting guerrilla style. Do you remember any interesting moments filming on the street?
I became friends with Eric very quickly and we did a number of things. Helene Winer was the director of Artists Space in its first manifestation, when Cindy Sherman worked behind the desk. She was giving out fifty dollars to anybody who wanted to make a film for a show. I took my fifty dollars and made *Waiting for the Wind*. Eric took his fifty dollars and we went out at night onto the Bowery, where quite often cars were abandoned, and there was a car which had been put up on blocks and had the wheels stolen. It wasn't going anywhere, and Eric wanted to do a film version of Warhol's paintings "The Car Crash." I went with him and Tina L'Hotsky to this car in the middle of the night and I got in the front seat and put my head through the broken window. Tina got in the back seat and put her head through the back window, kind of our whole bodies, and Eric went around with a bottle of ketchup and chocolate syrup and covered everybody. Then he got under the hood and covered himself in chocolate syrup. It was a crazy impossible car crash scene. There was no logic to the way people had ended up in this tableau. Michael [McClard] was shooting and he had a sun-gun and Super 8 camera. He circled this scene round and round, coming in closer and moving further away. We just froze in position and made this really intense and amazing film called *Car Crash*. I think [Eric] put sound over it later. The sound of helicopters, of chopper blades, which worked very well because it elevated this circling camera off the ground somehow. The thing I remember most was there was a bunch of homeless people on the Bowery. They came across us and were like, "Oh my God, this is awful. What happened?" They didn't notice that the car was up on blocks, and that there was a stink of ketchup. They were so appalled at this terrible scene. It was great.

With your own work you have these bigger narratives like *Rome '78*, but you also document performance and artwork. Did you approach the films differently?
Around the time I made *Rome '78*, that was a big leap for me because the most radical thing one could think of doing, within the world that we were working, was to make a narrative film. The art movie houses were like Anthology Film Archives. Millennium existed, but we were sort of anti them. We didn't want to show there because they were associated with another generation—people who were about two years older than us. But it seemed like a big gap. Which is why Eric, Becky Johnston, and I started The New Cinema as a place to showcase our films. So *Rome '78* was my first attempt at a narrative film. Looking at it now, it's not the narrative that's the glue that holds the film together, if there is any. I think there's quite a similarity between *Rome '78* and a film like *Street*, that I made much more recently, where I kind of set the camera and myself back from the action and observe. There's an active observation of things that I've set in motion. It was a kind of verité meets the imagination.

Is there a bit of that in your other films too like *Pendulum*?
Pendulum was a kind of documenting, but it was also a kind of animation. I was documenting this thing that I had made and set in motion. But it accrues menace as the film goes on. Now when I see it, the swing of the pendulum in the old streets of New York, it seems about the passing of time and things that weren't foremost in my thinking at the time I made the film.

A lot of Super 8, because it's made by artists on a low budget, ends up documenting a place and a period of time in a fascinating way which was never part of the intention.
Absolutely. Like home movies.

Can you describe The New Cinema scene?
The New Cinema was Eric's idea with Becky, and I was somehow in on the conversation. I was like the third astronaut, the guy who stayed in the capsule while the others went down to the moon. We rented a space on St. Mark's and we asked Michael Zilkha who had a record company, ZE Records. He had Suicide, Alan Vega on his label. He provided us with an Advent TV projector. We shot our films on Super 8 and then transferred them to 3/4" video at Manhattan Cable on Twenty-Third street with Jim Chladek. That was our version of having a print made. We showed the 3/4" videos on this Advent TV screen. You have a big projector which is like a chest of drawers and it sits in the middle of the room and has three big lenses, one blue, green, and red, and they project onto that big concave TV screen shaped thing. If you sit directly in front of it, you can sort of see it. If you are off to one side it disintegrates pretty quickly. But that was our movie theater.

How big was the screen?
It was like sixty inches by forty or something. Pretty small, but then the room wasn't very

INTERVIEW

The Contortions (L–R): Adele Bertei, James Nares, James Chance, Pat Place, and Reck.

big. We had fifty fold out chairs. And we wanted people to pay. We didn't want any freebies. We wanted to be legit. This was a real movie theater, and you had to pay two dollars to get in. But people happily paid their two dollars. It was a very smoky room with people standing against the walls and filling the chairs. We showed our films. We showed Vivienne Dick, Charlie Ahearn, Diego Cortez. The place stayed open for about seven months, and then we closed it, because we thought it was better to close it when it was doing really well.

Were you getting any kind of reaction from the people at Anthology or Millennium as being this kind of upstart?
Amy Taubin wrote that she thought that The New Cinema was a kiki scam to make money. Then she described the image on the screen as being like bent pink soup. That became a refrain for us. And she was exactly right, not about the scam to make money, but about the bent pink soup. It was an appalling image, but it was our image. It wasn't anybody else's and we were showing it the way we wanted to show it. We were dismissed and feared at the same time because we did represent a new energy. We had set ourselves up to be dismissed because we were being dismissing. Amy, of course, came around and wrote some great things about our films and has continued to support us and write about the things that we, as a body, have done.

When you all are making films did you have any thought about distribution or getting your films out to a broader audience?
Some of us were a little more savvy about that. I was never very savvy. I just found a letter from somebody from Paramount Pictures from that time saying, "I've heard great things about your film *Rome '78*. Would you be interested in doing this or doing that?" I didn't do anything about it. And really to a fault, I didn't know how to make the next step. Plus, I had retreated quite quickly from the whole idea of making any sort of narrative film. My next film was this video documentary with this IRA bomb maker. I called it *No Japs at my Funeral*, after something he says in the film. After that it was *Waiting for the Wind*, which is a kind of visionary film. More along the lines of Maya Deren.

Talk about restoring your films.
I've worked with Bill Seery at Mercer Media. Bill's wonderful with anything video because any kind of machine that could be useful he collects. He's figured out things like my reel-to-reel Portapak 1/2" tapes. He bakes them at a very low temperature for a couple of days and it reattaches the celluloid to whatever it is. The silver? And it solidifies without destroying the video. He has a good aesthetic sense and he's very knowledgeable. He's been scanning all these Super 8 films in 4K and it looks so good. I get a little more of the frame and the colors. And it's a question, as I look at these films and restore them, how close to stick with the original? It's a dilemma, but it's one that I resolved in my own mind. I want to see the film as I would have wanted it to have been seen.

Your intention rather than the execution.
Yes. I'm not so nostalgic with the patina of history—a little bit, but not in its fall apart way. So I take splices out most of the time. Sometimes I'll leave them in. Flash frames I always loved, especially when they were shot on that Nizo, because I used it purposefully. The camera had a pistol grip handle, that you pressed in, and it engaged the electrics, and then you could pull the trigger. If you released it from the electrics before you released the trigger, you had a good chance of having a flash frame because the camera would just stop. I enjoyed using that. Another thing about Super 8 I'm astonished at is the quality of the color on the Kodak film stock that we used. It's unbelievably pristine. The Ektachrome has kind of blued with age [but the Kodachrome] is incredible. ∎

INTERVIEW

Beth B

BETH B CAME TO NEW YORK in 1976 where she quickly fell in with filmmakers, musicians and artists interested in creating provocative work. Her films dripped with a social conscious, giving voice to the disturbances beneath the surface of American society. In this period, her partner in crime was Scott B, and between 1978 and 1981 they made some of the most challenging films to emerge out of the No Wave movement. B would go on to direct features, including *Salvation!* (1987), starring Exene Cervenka of the punk band X, and *Two Small Bodies* (1993). More recently she has embraced the immediacy of digital technologies, which harken back to her Super 8 roots. She is currently finishing up a documentary on musician/artist Lydia Lunch.

I first came across Beth and Scott B's films at the University of Michigan in 1986. I was already familiar with experimental film. That familiarity, however, didn't prepare me for the B's *Letters to Dad* (1979). The film features actors reading letters written to Jim Jones by his congregants. By the time I saw the film, the Jonestown massacre was already seven years in the rear-view mirror, yet the film still managed to be confrontational. The realization that the film had been made only one year after that event made it even more combative. I don't think I had ever seen anything quite like that, and the film left a mark. I moved to San Francisco several years later and discovered Naked Eye Video, the first video store that I had come across that catered to people interested in more than just Hollywood film. There, I found a tape of the films of Beth and Scott B. It contained the aforementioned *Letters to Dad*, as well as *G-Man* (1978) and *Black Box* (1978). It was one of the first tapes I rented. Though *Letters to Dad* had left a mark, I was still not prepared. *G-Man* and *Black Box* up the confrontation quotient. They pulsate and throb, and reek of desperation. Cops, dominatrixes, terrorist bombings, and torture are set against a decaying city. The soundtracks are loud and punishing. They are films that take no prisoners. These are movies whose goal is not to entertain, but to provoke and to question. I was transfixed.

What brought you to New York in the seventies and what inspired you to go down the artistic path?

I had been living in San Diego and then moved to go to school at the University of Irvine. I went to art school there for a couple of years and realized that I was speaking a different language than the people in Southern California. I never felt like I fit in. I could not find people who I could connect with artistically, intellectually. I couldn't find my tribe. I moved to New York with the intention of going to the School of Visual Arts and finished my degree there. Even before I finished my degree I was doing work that was on the streets. I started doing stencil messaging on the streets. My first stencil was "Where's the money?" and spray painted them all over the city. That was at a time when there was no money in New York. Everything had been closed up, people had been pushed out. The city was in a state of decay, burnt out and abandoned. In this wonderful way it became a playground for the disenfranchised. Everything that seemed impossible suddenly became possible because everyone was in that state of not having anything to lose. It was the beginning of punk, listening to a lot of music that was coming out of London, which was all about no future, but in a sense, we didn't even have a present. People were looking to recreate identities, to take the idealism that I certainly had and destroy everything around us that did not represent the reality of what was going on in this country, which was, in a way, truth telling.

My work has always been about truth telling. We grew up believing in the "happy family" and suddenly we were realizing it's all a fucking lie; it's all a facade. I found like-minded people who were interested in exploding all of the myths, even though we were completely unconscious of what we were doing and what we were feeling. We didn't have a clue except that we were disenfranchised. We were completely alienated from everything.

Were you feeling disenfranchised from the New York art scene that came before you?

Absolutely. Coming out of the School of Visual Arts, I already started to show in places like Artists Space and I actually had a couple of museum shows in Europe, but there weren't many places to show the kind of work that I was doing. For me, the most loathsome thing was going into the precious white spaces of the galleries and museums and sitting on my hands and being polite and seeing what I considered to be absolute garbage because it was primarily about form, it was about structure. It was, in my mind, masturbatory art that was not about content.

People were feeling this urgency to speak out about the injustices of our times and we didn't have a voice. People felt like we've got to do it however the fuck we can, with whatever means we have. And that created a surge of energy that led to a kind of collaboration, and that revolutionized the way that I was able to see the creative process. It was not about being a painter or being a sculptor or being a filmmaker or being a musician. It was about what was the essence, what was the message, what were you saying, what was the battle cry? Whatever form best expressed that, was what made the decision.

Because art was commodified and controlled, and the curator said what could come and what could go, it was geared toward people with money. And I had no money. Even before I came to New York, I was supporting myself by waitressing at a wonderful place called Magoo's which was where I met so many artists. In the kitchen, Arturo Vega was the cook. He would say to me, "I manage this really cool band called the Ramones," and I was like "Oh really?" And then suddenly I was seeing the Ramones at CBGB. You would just meet people on the streets and that's where the connections were made, where the creativity was coming from. So, it was a time of reinvention and saying, "We're going to take control and do whatever we want to do, and we will find the places to show it. We don't need you to validate us. We don't need to have your money or spaces. We will make the work, against all odds."

Did Super 8 become the perfect medium for this 'against all odds' mindset?
When I was studying art at the School of Visual Arts, I wanted my work to be accessible to anybody who came in from the street. Galleries did not have that feeling whereas clubs like Max's and CBGB and Hurrah did. You paid three dollars to go see a film or a band and anybody could walk into that club. That was appealing, moving away from the preciousness of the art world. It was a way of looking at the possibilities of bringing intense, uncensored ideas to the audience. What form would make it more accessible and what form could express all the different elements of creativity?

Super 8, by virtue of it being so small and so immediate, where you shoot it, you bring it to the lab, get it back the next day—that was so appealing. And using the medium with all the limitations that were inherent, which was you've got basically three minutes per reel. It was that kind of construct which seemed like a way to give the most freedom to merge different genres of expression in cinema. Super 8 had a non-precious quality that comes from a medium of home movies, not art, not commercial cinema. This somehow elicited a greater sense of freedom and experimentation merging documentary, narrative, and experimental modes. It didn't have the historical baggage and could in fact serve all of those genres because it was so light, and I could be completely mobile with it. It was not a big camera in someone's face.

The very first camera I used was a Kodak. I don't know which one, but it was an early Kodak box camera. No sound, no focus. You just pressed a button on the top and it filmed. For me, who knew nothing about film, it was perfect. Some of the first footage that I shot was included in the film *G-Man*. Some of it is on the streets of New York, documentary style. Some of it is sit-down interview style. Some of it is with actors, and some is very experimental. It was a way of deconstructing cinema, but still working within the narrative form. When I look back on it now, *G-Man* was one of the first hybrids. There was something exciting about using a medium that could also include that visual art aesthetic that I had conceptually.

For *Black Box*, artist John Ahearn painted all the backdrops. So, it was bringing art into the cinema. It was also bringing musicians into the films as actors and to create the soundtracks. These were musicians who had the same rebellious attitude toward music that we had toward art and cinema so sonically it merged seamlessly with the visuals and content.

***G-Man* and *Letters to Dad* respond to the political moment in a unique way. You use the immediacy of the medium to comment on what's happening in the streets of New York at that moment.**
Absolutely. It's about social politics. It was the failure of the hippies, the failure of post-war consciousness. There was no support for the disenfranchised, and yet there was a facade that there was. We were products of these broken families, of the abuse that nobody spoke about, about the authoritarian regimes that continued to commit atrocities and war. And yet nobody was talking about it.

In the seventies some radical groups started to scratch the surface with organizations like the Symbionese Liberation Front, the Red Army, and the FALN who were trying to explode that hierarchy. In some smaller and much tamer way, we were trying to destroy the hierarchy within our own lives. A lot of it had to do with the family, but also the politics at that time and the disproportionate wealth.

The streets were our playground where we could spew out all of that rage and alienation. We would break into abandoned buildings and film in them like, "OK this is our space now." We drove a car into the East River and it's like, "That's our East River." That was in the film *The Offenders*.

INTERVIEW

People were like, "Oh my god, you did that?" It was basically claiming New York as ours and saying we are not going to abide by the rules anymore. We're making our own rules. With Super 8, it fed into that kind of mentality because I didn't really know anything about film, and so it was making the films in the image of what I was going to construct rather than following any sort of formalized tradition of storytelling.

Of the No Wave films, I find yours the most cogent and easy to follow. Even though you are creating this new hybrid of storytelling, it was very controlled.
I think that's because it was. I do think both Scott and I were perhaps more politically minded than some of the other filmmakers at that time. Everyone from that time was lumped into the label, No Wave, but in fact, there were different aesthetics, intentions, and concerns. In order to make something commercial and marketable you have to give the impression that there is a larger movement than just the singular artist. And so, what became punk, what became no wave, new wave, were comprised of people doing really different things; but in the same time frame, in some of the same cities.

The films were quite unrelated in content, just as the music. Mars was four people playing on the same stage but as if they were in separate rooms, whereas with Lydia Lunch's Teenage Jesus and the Jerks, it was like she was slamming a nail into the coffin with deliberate social politics. Then The Lounge Lizards were this kind of sleazy lounge band. They were so different, and the filmmakers of that time were also different. But in order to commodify something and render it harmless—because people were very threatened by what we were doing—people label it, people put it in a nice little box so that it can be easy digested, written about by the press, and be commercialized.

For me there was a profound dissatisfaction with what I saw around me, and I was politically activated. It had to do with certain disturbances that I grew up within my household and the politics of the male hierarchy. It was so present in my household as a child, that for me and for other people I know, especially for Lydia Lunch, the work had to be about voicing what we as women were battling from a young age, and that we weren't going to fucking take it anymore. People needed to hear our stories. If you notice in most of the films that Scott and I made together, there are powerful female protagonists. No one was going to stop me from saying what I needed to say, but it was prescient in that no one was talking about the domestic and governmental abuses at this time. Yes, there was feminism and the radical feminists, but they were not talking about the abuses, they were talking about rights. They laid the groundwork really profoundly for our generation, but they were still unwilling to uncover what was happening in the homes.

In that moment in New York there are a lot of strong women artists—yourself, Lydia, Vivienne Dick, Becky Johnston. Is that a good group of people to have, so you're not doing what you're doing alone in the wilderness?
There was no consciousness of it. I mean we didn't know what we were doing. We didn't have a clue. We were like deer in the headlights thinking we had to fucking smash into those headlights, but we didn't know why. We were going to smash it all down. It was just knowing that what we had experienced was not right and a feeling of us versus them mentality.

I was actually not close friends with Vivienne, although deep respect for her films and now a dear friend. I was more involved with some of the visual artists like Tom Otterness or Coleen Fitzgibbon or filmmaker Richard Kern. It was more based on politics than there's a woman over there and she's doing cool work. I felt more aligned with artists, filmmakers, musicians, creative people, theater people who were dealing with difficult content.

I loved the gritty economy of Super 8, which seemed synonymous to the landscape around us. What was amazing about the filmmaking at that time was the Ektachrome blue, black, and red tones that were such a wonderful visual reflection of that content.

What destroyed that whole movement was the media discovered us. I can say gratefully Film Forum showed *The Offenders* after it showed at Club 57, but in a way that was the downfall of the movement because then people came in and said, "Oh yeah we'll give you a little money to do this" or "We'll give you a grant to do that." Then the focus digressed to money and success, which led to competition.

There was a review of *The Offenders* by Janet Maslin in *The New York Times*. She was saying, "And these characters are wearing

---

"The work had to be about voicing what we as women were battling from a young age, and that we weren't going to fucking take it anymore."

leather jackets" and she was really disturbed by this leather jacket wearing. Then of course, in a few months Bloomingdale's is selling the same leather jackets and Madonna is sporting leather. What comes from the streets finds its way into popular culture, and then the popular culture takes it and it is no longer the primal scream. It becomes fashionable and commodified.

What was the path for you like after that era?
I went from a wildly prolific time of independent filmmaking to not being able to get my films made in the nineties. So, to survive, I went into television from 2000 to about 2010. I made a lot of great documentary films for some of the networks, but coming out of that was like reliving the kind of urgency I felt in the late seventies. Small video cameras were coming out where you could just put a microphone in the shoe mount, and there I was going back to my roots in Super 8 filmmaking. Uncensored, a fuck you attitude, doing it all on my own without a crew.

I made Exposed in 2013, which brought me back to my stomping grounds in the New York underground. I filmed most of it myself with a small camera. That liberation to do it my way was like a return to Super 8. Now, I'm working on a feature documentary about legendary performer/musician Lydia Lunch and I find that the spirit and language of Super 8 is very alive in this film. That's how I've made three feature films in the past five years—going back to shooting, editing, producing, directing them myself.

Lydia Lunch and Bob Mason in Black Box *(1979).*

That was always the promise of Super 8. That ease, that immediacy. I feel like we're getting there now with technology today.
I feel like it absolutely fulfilled that promise for me. I felt I did not need a crew. I did not need all of those things that people often made as excuses for not producing work. And people use it today and I'm like, "Oh, come on, don't give me any excuses, just do it." There is no excuse. If you have something to say, then you will find a way to voice it. The most important thing is to be able to voice what is not being heard, what is not being said. My work has been about that from day one, and it remains about that today.

You can always find a way, even if you have a just a single penny, because you can always find like-minded people and collaborators if you are willing to open yourself up to the concept of collaboration. Film is inherently collaborative, it just depends on you as a director, how open you are to hearing others, and working your fucking ass off. That's the other thing that I became humbled by in my later years. I think I'm pretty damn smart and creative, but you know what? There are some brilliant people out there and when they are willing to offer feedback and words of wisdom, I'm grateful to listen to them because I'm not always right. I've gotten extraordinary enlightened ideas from others that I then apply to my work. I want my work to be the best that it can be, and sometimes I'm too close to it to know what that is. When I was young the concept of being vulnerable was frightening to me. I was scared even though I was doing ballsy work. I was scared because I didn't quite know what I was doing and didn't have the confidence I have today. But now I know that being vulnerable and being open to hearing other people's opinions can be a great gift. In those days I didn't understand that at all, and in some ways, I think it was important for me as a young artist not listen to other people because I had to discover my aesthetic, find out what was important to me and make my own mistakes. Today I know that hearing those other people makes my journey in life more powerful. ■

INTERVIEW

Narcisa Hirsch

NARCISA HIRSCH IS an Argentinean filmmaker who started making Super 8 and 16mm work in the late 1960s in Buenos Aires. She was a core member of a performance group that created street theater "happenings," and is best known for *La Marabunta*, a giant female skeleton covered in food and filled with pigeons. Hirsch became interested in filmmaking while documenting the public's reaction to this interactive performance piece. She worked in Super 8 throughout the 1970s, but shifted to video when it became more widely available. There has been a recent revival of her work, in part sparked by the *Ism, Ism, Ism: Experimental Cinema in Latin America* exhibition at the Los Angeles Film Forum in 2017.

How did you get interested in filmmaking?
I was a painter. We had an art critic here in Buenos Aires that was very prominent, and he said that traditional painting was dead, and what was happening were "the happenings." What you call performance now was called "happenings," which is what people did mainly in the States and in New York. I had two friends, and with them, I started doing happenings. These were things like we were giving away apples on the streets like a gift, for example. We took photos of that and filmed it in Super 8. Then I had a big happening which was called *Marabunta*, which was a big female skeleton that was filled with food and and live pigeons. That was with the premier of the film *Blow Up* by Antonioni. This was '67. *Marabunta* had somebody filming it in 16mm. We went to edit it and I started getting interested in the movement, and in doing something which was not painting, but was still in the field of the visual arts.

When you say the movement, is it a political movement or an art movement?
Movement in the sense that, obviously, painting is not something that moves. In the real sense, in the pragmatic sense.

So, what makes you pick up Super 8 as opposed to working with 16 or video?
In the late sixties and early seventies there was only 16mm, Super 8 or Regular 8. In the early seventies I got interested in film. I had a few short films in 16mm because my husband had bought, for family reasons, a 16mm Bolex camera that you wind up manually. I used that camera. That had become too expensive to do as a hobby. People were using Super 8 for their traveling or for the family birthdays and fiestas. This is why Super 8 came up. Then a small group of people got together who were doing short films in Super 8 which were not for family reasons, which was film. At that time, we had very little contact with others abroad because it wasn't like it is now that you can get a DVD. You had to go to New York or San Francisco on the West Coast where things were happening and see it there. So this is what I started doing. I went to New York and I got in contact with people who were showing experimental cinema. I had contact with people who would tell me what one could see and hear. There was also the sound—if you like Steve Reich and these musicians of the seventies.

Did you work with Steve Reich on one film?
I made a film with the sound of Steve Reich. This is a film that has become fairly well known. It's called *Come Out and Show Them*. I showed him the film in New York. He didn't like it at all. He was very worried—who was going to buy it and he wouldn't get the money for it or something like that. [Laughs] And then I had made [the song] shorter. My film was ten minutes long and his work is twelve minutes long. He was angry that I had taken out two minutes. So that was Steve Reich.

So, you're with this group of people who begin making Super 8 films. Are you self-taught? Are people going to university to learn this? How is the Super 8 scene coming together in Buenos Aires?
It would happen mainly in one's own place, in one's own house. There were a few places like bookstores or cellars that one would put together and show what one had done. There were a few others who also begun to make films in Super 8 and they would appear and show their stuff. We became like a group. Very few of us, like six people maybe. And to see each other and show our films to each other and a few friends. There were never more than ten people.

Can you talk about the political moment in Argentina? Was that influencing your work in some way?
In the late sixties and the early seventies there was a lot of upheaval in Latin America.

It also was happening in the States. These were the times of the civil rights movement, it was the time of the Vietnam War. The Americas, as a whole, were much influenced and under the domain of these political happenings. At that time, you either were the political right or you were on the political left. You couldn't just not be anywhere. Experimental film was a place where one could sort of hide away from the political movements. Art, especially modern art, is undecipherable, especially for those who were in politics. I mean, if they would see a film like *Come Out and Show Them*, it wouldn't mean anything to them. So this is where I was, and this is where most of the people who did what I did were.

When do you start exhibiting your films internationally, outside of that small group of people you're working with?
In the eighties, when video came up, all the avant-garde, which was so strong in the United States, people like the Factory of Andy Warhol, all those people had either disappeared or were not doing the avant-garde anymore. It had become pretty flat. The people who were doing these avant-garde films would start doing video, and that was a different language. I have a small group of filmmakers who come together and we see films that are not shown here commercially. These people are young, they are thirty or so, and they hadn't been born when I was doing my films. They probably feel that when we were doing our films that things were happening, and that nowadays nothing much is happening. So they're looking our way and they want to see the Super 8, and want to see what we did in Super 8. The young people that go to film school, they want to film in Super 8, and buy cameras and projectors. I think that the new generation is curious about what had happened at the time. [There] is a certain curiosity for the old times because of its content—its revolutionary, avant-garde content. That is what has made me more well-known. Now there's a festival in every little town, and you have to feed these festivals, no?

Is your film *Portraits* Super 8 or 16?
There are several *Portraits*. I have *Portraits* done in 16mm and I have *Portraits* in Super 8. I had a friend who was a psychologist, psychoanalyst—a woman. We had, at the time, groups of women coming together and talking about gender, like it's happening now. That started in the seventies. My friend said, "Since you are a filmmaker, why don't you make a film of each of the women who were part of the group? Why don't you let these women talk to their own image?" [This] is something that she invented. I think that was very effective because I filmed these women with a still camera with no acting, no speaking, no nothing. Three minutes. Then I would project these images to that woman, and each of these women who would see herself projected would then talk to herself. Speak and have a dialogue with herself. That was the sound of the film. That is a film I did several times with several different groups of women. I myself appear in three different years with the difference of two or three years in between. That is one film which is called *Woman Talking to Herself*.

Can you talk a little bit about *Bebes*?
As I told you there were these happenings, and one of the happenings I did was buying baby dolls for children to play with. I had bought a big amount of these baby dolls, and I gave them away on the street in Buenos Aires, in New York, and in London—the same babies. I myself alone standing on the street dressed the same way in the three cities and saying, "Have a baby." That was one happening, and when that was finished, I made a film. There was a 16mm version and a Super 8 version. The important one is the 16mm one. It's a longer film called *Pink Freud*.

When did you move on from Super 8?
Personally, when Super 8 had died physically, I went back to 16mm. I started doing a few films, and even a feature film. I said, "Well, I'll see what happens if I do a real film with actors." I did that and I realized this was not my cup of tea. Then video started to be more present. I met a woman who was professionally editing videos. She came to work with me, and she has been with me now for fifteen years. She does the editing for me. Now I don't really film very much. I use what I have, like the films that you throw away, the footage that you don't use anymore. I have a lot of that and I use it. I transfer it to video or I just steal other films on Google or wherever one can steal. ■

> **"There were a few places like bookstores or cellars that one would put together and show what one had done. There were a few others who also begun to make films in Super 8 and they would appear and show their stuff. We became like a group."**

THE 1980S AND 1990S

In the 1970s, the possibilities for Super 8 seemed endless.

Then video arrived and Super 8 was a medium on the ropes. However, there was still gear to be found, mostly used, but still functional. University programs continued to offer Super 8 in introductory courses. Throughout the 1980s, you could find Super 8 being taught at University of Michigan, UCLA, Brown, the University of Miami, Massachusetts College of Art and Design, Temple, the University of South Carolina, San Francisco State, and more. Though it was seen as a starter medium, many students used it beyond their first-level classes.

College curriculums introduced students to classic experimental films by the likes of Maya Deren, Stan Brakhage, and Michael Snow. Not surprisingly, Super 8 filmmakers began making art films in this vein. You can see Brakhage's influence on Paul Clipson and the collective, silt, who use the film strip as their canvas, creating arrestingly beautiful work. Found footage and collage filmmaking, stylistic staples of the experimental world, were finding new converts in Super 8. The influence of Bruce Conner can be seen in the works of German filmmaker, Matthias Müller and Luther Price. One can see the impact of Larry Jordan's animated collages on Super 8 animators Lewis Klahr and Martha Colburn. Super 8 filmmaker Peggy Ahwesh studied with experimental icons Paul Sharits and Janis Crystal Lipzin, her films mixing elements of narrative, documentary, and performance in a potent experimental brew.

While more accomplished art films were being produced in the 1980s and 1990s, Super 8 still had a firm grip on the imaginations of filmmakers interested in telling stories. Richard Linklater (*Slacker, Dazed and Confused*) shot his first feature on Super 8. Jim Sikora made skilled Super 8 narratives focused on the hardscrabble world of blue-collar Chicago. Dave Markey created no-budget narratives starring the denizens of the Los Angeles punk scene.

Super 8 continued to have a reputation for being raw. This was in part a result of so many Super 8 films being made by first time filmmakers, but it's also a part of the no budget DIY aesthetic that grew up around the medium. Super 8 filmmakers learned their lessons on the job and accepted the fact that mistakes would be made along the way. Super 8 filmmakers weren't afraid to include their blunders in the final film, after all, most of them couldn't afford to reshoot. They reveled in the rawness. This made Super 8 the ideal film gauge to channel the punk rock vibe. Jim Jarmusch (*Down by Law, Year of the Horse*) came of age during the No Wave film moment in New York. He sums things up quite nicely. "Rock 'n' roll bands said, 'Fuck virtuosity. We have something that we feel, and even if our expression of it is musically amateurish, it doesn't mean that our vision is.' That helped me, and other people, to realize that even if we didn't have the budget or the production structure to make films, we could still make them using Super 8 and 16mm."

This attitude carried into the 1980s and 1990s. G. B. Jones and Bruce LaBruce picked up Super 8 because that's what they were handed in college in Toronto. They stuck with it in large part because it fit their no-budget lifestyle—and they were technophobes. But they weren't going to let the technology get in the way of saying what needed to be said. LaBruce, Jones, Markey, Colburn, and Ahwesh were all informed by this punk rock spirit. An imperfect medium, Super 8 was quick and dirty. Tape splices were visible, sound could drop out. These types of imperfections added to the energy fueling many films in this time period.

Enthusiasm for Super 8 continued into the 1990s, aided by an explosion of underground film festivals from New York to Honolulu. Operating from North Carolina, Norwood Cheek

popularized the Flicker series of monthly screenings with no submission fees. Incredibly popular, Flicker screenings popped up in over a dozen cities. Melinda Stone started her itinerant Super Super 8 Festival, taking the show on the road and travelling to towns across the country. At each stop, Stone teamed up with local bands to accompany many of the silent films she programmed.

From a technical perspective, Super 8 was in retreat. Most equipment was no longer manufactured, sound stocks were discontinued, as were beloved film stocks like 4-X. Yet in spite of it all, the 1980s and 1990s were vital decades for the medium. The following interviews capture that enthusiasm, and reflect a medium that was no longer in its infancy, but settling into a more comfortable middle-age. ∎

INTERVIEW

Richard Linklater

RICHARD LINKLATER IS one of the most acclaimed directors of his generation. His films include *Dazed and Confused, Before Sunrise, School of Rock* and *Boyhood*, which earned him Best Director, Best Picture and Best Screenplay Academy Award nominations in 2015. The film that put him on the map was *Slacker* (1990), a low-budget, 16mm romp through Austin, Texas, following the ramblings and musings of a contingent of college-town misfits and weirdos. Though it was shot on 16, *Slacker*'s final scene is a teeny Super 8 movie unto itself, which culminates with a Super 8 camera being chucked off the side of a mountain. *Slacker* began Linklater's longtime collaboration with friend and cinematographer Lee Daniel. Prior to *Slacker*, Linklater wrote, directed, produced, edited, and shot *It's Impossible to Learn to Plow by Reading Books* (1988), a no-budget feature on Super 8. A minimalist road movie, *Plow*, follows an unnamed lead as he skips from town to town, meeting and hanging out with friends. Before he started making films, Linklater worked on an oil rig, saving his money to buy top-of-the-line Super 8 equipment on which he could hone his craft. He never went to film school; he learned by reading books about film technique and put those lessons into practice on his Super 8 gear. Additionally, Linklater founded the Austin Film Society in his house in 1985.

How did you get interested in filmmaking?
It sort of found me at age twenty or so. When I grew up in East Texas, our generation, no one knew what a film director did. Films were just these magical things that showed up from Hollywood and you liked them, or you didn't. But no one really knew much about the process. The culture didn't emphasize it. The only filmmaker I knew growing up was Alfred Hitchcock. He was famous. He had that TV show, and I still didn't know what he did. That was not on my menu of artistic expression for the longest time, although as a junior high kid, my cousin and I, over one summer, shot a bunch of Super 8. I got good at camera tricks. I explored the medium and really liked it. I loved getting the film back. I liked watching what we shot. I liked editing. I liked doing in-camera edits. It deposited somewhere in my brain that, "Hey, I like shooting film." I got a taste of it. I knew about it, but it was on hold. I said, "Someday I might pursue that." I didn't know where or when, but when I did start to think of film as my medium, I started thinking I can write for film. I was always a writer, studying English and theater and, in college, writing plays. I started seeing some films that I loved. Seeing a few indie films that were done inexpensively made it feel closer to home. This was the early eighties. I remember *Chan Is Missing* by Wayne Wang, *Return of the Secaucus Seven* [by] John Sayles. And then a lot of films that are completely forgotten, maybe they're not that great, but they're inspiring. That person got a camera and went out and made a film. I was, at this point, working offshore, but saving my money. I'd take a week off work, I'd go to the Houston International Film Festival, which is a really nothing film festival, but I loved it. I would watch films of all kinds. I watched three films a day. I bought a book on the technical aspects of filmmaking. I was working, and I said, "At some point when this is over I'm going to buy equipment and go that direction." After two and a half years of working, I'd saved up some money. As soon as that was over, I started cracking the book, starting to read [about] every little element of filmmaking. By that point, I had friends who had been through film school. I'd done enough research, I'd seen some interesting Super 8 films. There's one made in Austin called *The Invasion of the Aluminum People* by David Boone. There was another one called *Speed of Light* by Brian Hansen. That's 16mm. Those two showed at a gallery in Houston while I was working there. That really influenced me. A–Pick up a camera and do it. B–it should be a Super 8 camera. And these films came from Austin, so maybe there's some kind of film community there. I never really found much on my own. I had to reinvent or try to create a new film community. Lee Daniel was around, and I got to know him. He was showing a student film he had made, and we realized we both liked Super 8. I showed him my house and it was like a Super 8 studio. I had the best Super 8 projector, I had a great camera, a lot of film stock. [I had an] Elmo 1200 projector, which is good for sound if you want to record sound. I had a Canon 1014 [camera].

You had the best stuff.
Well, I had $18,000 in my pocket and I was serious. I got a good Bogen tripod, got good sound equipment. I jumped

The Super 8 scene in Slacker *(1990).*

INTERVIEW

in, man. I was like, "Okay, I'm a filmmaker now. I'm going to teach myself everything." I sat down for a couple of years, and since I wasn't working, I wasn't in school—eighteen hours a day of film school of my own making. I would shoot short films, edit all night. Always mailing off film. It was fun to be so excited about the medium. I'd do entire films as lighting exercises, just getting shit out of my system. They were technical exercises largely, but at some point along the way it all culminated. I made a Super 8 feature before I did *Slacker*. I spent two years on a Super 8 feature. That was my grad thesis film.

Was Lee at school at University of Texas?
Yep. I was never enrolled but I was kind of around.

Did you feel like you didn't need to go to school and you could figure it out on your own?
Yeah. That's my personality. I don't like authority. I would audit classes. I did wander into a few film classes and I've told this story a bunch because it was a turning point. They were showing some films and the teacher was like, "Where's the close up of the hands?!" I was like, "Oh, God. I don't want somebody in my ear telling me, 'Where's the close up?' I don't want to do a close up there. I don't want to have to do anything. I didn't like the idea that there were rules, that you're being taught a system. That made no sense to me. That whole thing about, "You need to know the rules before you can break the rules." Yeah. But the rules are pretty obvious. Just watch a few films. I knew I wouldn't be able to communicate what I had in my head cinematically because I had some really crazy fucking ideas about film. I put a lot of theory and thought into what I thought could work in a film. I had a lot of ideas I wanted to explore. I didn't want a sounding board. I didn't want an authority figure telling me what I'm doing wrong. I didn't want to be judged.

When you made *It's Impossible to Learn to Plow*, what was your vision of that movie? That's a movie that does break a lot of rules. I don't know that you ever go to the close up of the hands.
I was under the influence of seventies structuralism. Minimal. Chantal Akerman, James Benning, Jon Jost, Jim Jarmusch. I thought that's an interesting way to tell a story. In the eighties we were being polluted with MTV imagery. Cutting had lost all meaning. It had depreciated the value of the image or the cut because it was just constant. I was thinking the most radical thing you could do—André Bazin mise-en-scène, let the viewer decide what to look at. That kind of thinking made sense to me. Also, on a practical level, when you're doing a film with no crew or money, it's like, "I don't need a script supervisor to tell me which hand the glass was in." I wasn't trying to tell a conventional story. That film rebels against everything. I don't want the camera to move. I wanted there to be a lot of camera movement within the frame—people walking or [shooting] out a window of a moving train. There's a lot of movement, it's a road movie, but I never wanted the camera to move. I wanted every shot to be a composed portrait. I had a theory behind it all. It was definitely in the minimalist, oblique narrative tradition. I liked the idea of an unnamed lead character. It's youthful alienation. I wanted a film that was itself alienating on all levels. It felt to me the right form to meet the content. In my next film, *Slacker*, I would flip those things on their head. I wanted constant camera movement and constant verbal.

You're now talking about *Slacker*, was that decision to flip the script a response to your first film or was it that you now wanted to try something new?
Let's just say I got that out of my system. I dug a little deeper into myself. How does anyone find their voice in this world whether you're a writer or a painter? You have to work through a bunch of stuff. It takes a while. I had the idea for *Slacker* five years before I started shooting. I had been thinking of that narrative structure, but when I found the visual corollary, it was the opposite. Suppose that character in *Plow*, that nameless guy, opened his mouth and started talking. And once I did that, I realized I had all these words in me—the playwright in me, the short story writer. That was my pre-film origins. I just had gone down a more experimental, structuralist [direction]. *Slacker* to me, it's what that film needed to be. I had the language to tell that story. The camera always moves, but we're still not cutting to the close up of the hands. It's striving to be one take, one scene like my previous

> "Super 8, if you're doing sound, you get your two and a half minutes. It's all important. It's costing you money. You got to have a plan."

film, just there's a lot more ground covered and movement and energy and humor. But I still was rebelling against traditional film syntax, notions of master shot, close up, medium shot, shot reverse shot. That all was ugly to me. That would come later. I would have to slowly, begrudgingly tell a story that that would be effective for. That's what a director does. We all have stories, but what should it look like? What's the form to meet your content? I always try to go into every film with a strong visual theory and plan. That's where your collaborators are important. Lee got what I was trying because he's not a narrative storytelling guy. He's more of a documentarian, experimental guy. That's what we bonded over. I did *Plow* alone. The first time I worked with Lee was on *Slacker*. We really did see it like-mindedly. I didn't have to explain. I had other crew people going, "Just do a cut there and then turn around." I'm like, "No, I don't want to." [The crew would say], "It would be so much easier. Why are you doing all these dolly shots? Just cut." I was like, "No, I don't want to do that." Lee wasn't the one saying, "Let's cut." He got it. He's like, "I'm up for that challenge. I don't want to do a cut."

What was that like for you, jumping from 8 to 16 mm, and working with a crew.
[*Plow*] was a private film. It was a huge leap. You would think the next one, going from *Slacker* to *Dazed and Confused* was the big leap—an indie film to a studio film. But psychologically the biggest leap I ever took was—again, I was not that extroverted, and to ask five to seven other people to give me their time to do this crazy film I've been thinking about forever. Then to run a crew. All the leadership things you'd have to do, all the confidence you have to have. It's a real test. It was huge to suddenly have production meetings with people and plan this out. It was a rush to look up and realize I was getting this film made and these people were committed to working on it. Communicating your ideas to someone is hard. It's hard to be confident enough to throw down and say, "Okay, here's what we're doing." I realized this is my biggest test to date. The main thing is, forget whether you're good or bad at it, do you like it? Do you feel comfortable? For me, I realized there's no place I'm more comfortable than on a movie set. It was like, "I'm completely made for this. This is where I'm meant to be."

I am proud in *Slacker* there is a Super 8 section at the end that I blew up to 16mm. A friend was in town and she had a JK optical printer. I charmed her into showing me how to do it. I would stay up all night optically printing to 16mm. That blow up is homemade for the very last sequence in the movie. That in itself is a little experimental film, the kind myself and Lee were making. Two cameras, people just filming themselves. It had a fun energy.

In that scene you throw the camera off the side of a mountain. Did you actually retrieve it? Is the footage that we see at the very end of the film, the actual film from the Super 8 camera that went flying over the edge?
A little bit of it. The very first, like, strobe. But then the spinning part came from some footage Lee and I had shot at a party several years before. We put a Super 8 camera on the end of a fishing line and cast it out. I love that physicality. I put a camera on the ground and back a car over it. Not run it over. It was like a musical instrument. You play it. You play catch with it. Throw it off a thing. In Austin, Mount Bonnell is a destination. You hike up to the top, watch the sunset. Our little band of young people they run up there filming each other, and then the last guy throws the camera off. We filmed him doing that, so there was a camera down there. We taped that camera up pretty well, but in hiking down there, dammit if it wasn't still running. The film had run out, but it was still purring along. It was on, because we had to lock it on, and it was the sound that helped me when we were looking for the camera. We heard [makes the sound of a whirring camera]. Isn't that funny? I'll never forget it. The footage was a little too crazy, but we used a little bit of it.

When you made *Plow* were you thinking it could get distributed to festivals?
I never submitted that to one festival. I made VHS copies, I showed it on our local access station. I had to because I had used their equipment, so I owed it to them to show it. But that was about it. I wasn't in that frame of mind. I knew what it was and what it wasn't. I knew the world didn't give a shit. I was not that delusional. But I knew what it meant to me. It was part of my own development. Looking back, all those people working with me on *Slacker*, it wasn't just a guy who had a film, it's like, "This guy made a Super 8 feature." Everybody's a wannabe filmmaker at this point, but I had actually done whatever you have to do to conceive, shoot, edit and declare finish. That definitely put me up a notch. People working with you, they want to know that what they're working on will have a final form. I wanted to be a finisher. I'd met enough people that worked on their film for ten years and hadn't finished. My general outlook is I have a lot of films I want to do. I don't want to get bogged down with one. That's what *Plow* was. I'd finished it. By the time I was done with it I'd moved on. I was ready for for what was next. It never really showed, but it has a nice little home now as an extra on the *Slacker* Criterion. It's only 30 percent embarrassing as a piece of juvenilia.

INTERVIEW

Peggy Ahwesh

PEGGY AHWESH MADE an impressive batch of Super 8 films in the 1980s. She went to college at Antioch where her professors included legendary experimental film stalwarts Tony Conrad, Paul Sharits, and Janis Crystal Lipzin. Post college, Ahwesh returned to her native Pennsylvania, where she became the programmer of Pittsburgh Filmmakers, one of the Media Arts Centers that played an important role in the creation and exhibition of alternative media in the pre-internet age. Back in Pittsburgh, Ahwesh surrounded herself with like-minded outsiders, artists, and musicians, and began to create her own singular brand of Super 8 film. Mixing doses of narrative, documentary, performance, and improvisation, Ahwesh captures the vibe, lifestyles, and philosophical musings of her friends and collaborators in films such as *Philosophy in the Bedroom* and the *Pittsburgh Trilogy* series. Like so many Super 8 films of this period, Ahwesh's create portraits of people and places whose voices would not otherwise be heard. Ahwesh moved to New York in 1982. Though she isn't working in Super 8 anymore, she remains a force in the experimental film scene and teaches in the film and electronic arts program at Bard College.

What drew you to film as a way to express yourself?
I was interested in the arts when I went to Antioch College in the seventies and studied with Tony Conrad, Paul Sharits, and Janis Crystal Lipzin. Those were my main teachers. Tony, who is kind of a do-it-yourself guy, was always like, "Your daily life is part of your art." Your daily life was sound and picture and you were moving and performing or making music. I quickly moved from photography to media to film really easily.

That's an incredible batch of teachers.
Really influential on me, especially Tony [who] presented himself as this example of how to be an artist. I went for it. I remember being seventeen and being in his place and we were making music with kitchen stuff, banging on pots and pans and somebody was filming it. That's just kind of what you did. I was also very interested in music and performance.

Why did you start making films in Super 8 as opposed to 16mm or video?
After college I went back to Pittsburgh, where I'm from, and got involved in the film community there. I was the programmer at Pittsburgh Filmmakers in the early eighties. There was not a lot of money. It was not a professional geared crowd. It was more of a party and lifestyle group, and Super

What were the things you loved about Super 8, and what were some of the things that frustrated you?
When you talk about Super 8 you just have to enjoy what it is, not what it isn't. I love Super 8 to this day. Kodachrome 40 was the most beautiful film stock ever invented. The ease of operation, the camera's mobility. Ultimately though, you're dealing with a small little frame and it's a more volatile gauge because of its size. But I did and do love it in theory and practice. Those who know it feel that way too, and it's consumer based. When I first started, I could get Plus-X for $4.99, stock and processing. That's unbelievable. Looking back, that was nothing. So, it was affordable. You could buy it as a consumer product anywhere, and there'd be specials. There was this film lab in Ohio or somewhere. I used them pretty exclusively because they were cheap. Years later someone told me, "Oh, that's a terrible lab." I'm like, "Oh, it was the only one I knew." Again, a consumer product. You're a sitting duck consumer stuck with whatever comes your way. But that had its own charms. Every now and then you'd get something weird, some chemical. That's why it felt volatile. I'm grateful I came up in the Super 8-16 era. Splicing film and doing everything you had to do. It's a precious resource. You only have so many minutes. Super 8, if you're doing sound, you get your two and a half minutes. It's all important. It's costing you money. You got to have a plan. It organizes your thinking. It imposes a certain discipline, not to mention the obvious photographic element. It's film—you got to know your shit. ∎

INTERVIEW

Jennifer Montgomery in Martina's Playhouse *(1989).*

8, the portability of it, the invisibleness of the camera, the unthreatening-ness of the camera, the ability to shoot a lot and not worry too much about making a product—all those factors added up to Super 8. I liked the look of it. I had two matching Elmo's. They both took 200 foot magazines. I had bought them from a news camera guy who'd gone to Poland to document the Solidarity movement in the seventies. To me they were high-end cameras, but I got them cheap. So I had these two matching sound cameras with mag stripe. I used them for years. I still have them. And I had an [Elmo] ST 1200 [projector] and then a GS 1200 and I had a GOKO Editor.

At the time in Pittsburgh there were many other Super 8 filmmakers and we banded together in a really great way. We would go out and shoot a lot. It wasn't script-based and production-based like a traditional movie where you'd have a producer or a script and you'd shoot with a crew. We all had cameras and we were all out shooting everything. I would go to a picnic and shoot or there'd be a parade or I would shoot some drama that I had perhaps set up in my living room. It was very Warholian.

How do you think Super 8 helped define your aesthetic, particularly in some of those early films?
I was into odd characters. I collected people, and the sound capacity of the Super 8 allowed for a certain confessional modality. By the time I got to New York in the mid-eighties people were coming to me and asking to be in my movies because they wanted to tell me some story, they wanted to confess something. They didn't know why exactly, but Super 8 allowed for a transparency. Maybe video does now. People don't freak out so much when you're shooting something with your phone. But if you had a 35mm camera with a crew and a lot of lights, that would be a whole different thing. I was interested in that slipperiness between home movies, documentary, and art films, and how you might get that. Super 8 to me was an ultimate tool for being able to enact those modes without a lot of obstruction. And I like improvisation. I'm from Pittsburgh. We're descendants of Warhol. Long takes of people talking is very interesting. Crazy characters, people telling confessions and stories. Small setups in rooms where people just do something. It's very much a Warhol tradition. I was interested in Jack Smith and Kuchar and other people that work this way. I was interested in playful narratives, people, portraiture, things you could capture while you're moving around extremely portable. I had this simple setup with a lavalier on a cable and I would actually attach it to people. I'd have a really long cable and they just walked around with their mic attached to them. The whole thing ran on batteries and I would just pop in new cartridges as I went along. I also liked the look. It was soft and immediate in a way that I thought was graceful and a very lovely interpretation of the world. I wanted the world to be interpreted. I didn't want it to be factual and normal. Whatever the contemporary technology seems to be makes the world look normal, and I was interested in some other interpretation that was more poetic. And I did a lot of hand processing back in those days. The chemistry gave it another level of poetic intervention.

What made you choose hand processing?
I've always been interested in getting past the normal visual field into some other dimension. My interests are somewhat quotidian about people and place and

INTERVIEW

situations, but image-wise I want more of a layer to another dimension. So the [hand] processing is one of those things. I wasn't interested in nostalgia like how a lot of people use hand processing now. I was more interested in the materiality of the strip and how it could make a veil to another another place, another dimension, another reality.

You're interpreting the world and you're creating an aesthetic that is going to add to that.
Exactly. I was in the Biennial with this movie, *The Color of Love*, which is from the nineties. That's a film that was a Regular 8 porno film that I blew up on the optical printer. That film had all this decay in it because the film got rained on, so the emulsion was starting to go. It had this weird effect. I loved the effect, but it was a mistake to think that I liked that because it was nostalgic for old-timey film effect or to make it something like a painting. Although it did that, I was interested more in the decay of the material. It was about its impermanence and its historical fading. Someone sent to me the other day a page with a strip from the film and then across from it was this quote from Karl Marx, "In history as in nature, decay is the laboratory of life." I was like wow. "History and nature, decay and laboratory. I love all those words."

Can you talk more about Pittsburgh Filmmakers?
It was this charming little place in Oakland, which was a college center near the university. You'd go up the stairs, it was on the second floor, and at the top of the stairs, was that image of Maya Deren with her hands up as if she's looking through the window. If you went to the right, there was a screening room, and if you went to the left there was a gallery. There was a school in the back. They taught photo and film. There was a photo lab and there was an equipment office. It was a place to go at the time. One time I went there and David Cronenberg was there. He was in town and just showed up because it was a film place. The history of Pittsburgh Filmmakers precedes me by a decade. They had started in '69 or '70 maybe [Editor's note: 1971]. It was already established with grants and a regular calendar and students from the university took their media classes there. It was a stop on the circuit. People would write to me because they were coming through and they'd like a gig.

How did you get your films out and start making a name for yourself as an artist?
I was the programmer at Pittsburgh Filmmakers, so there were a lot of connections in various cities. I knew a lot of people. I also programmed people in Pittsburgh onto my calendar. That was a big thing. That was like, "Come on you guys, you've got to get your films ready because your show's coming up and you're on the calendar." That worked that way. I came to New York. I had shows at the Collective for Living Cinema and I did shows at Millennium. At the time there were lots of screening rooms around. I went to Philly, I went to Cleveland. This was a network of screening rooms that doesn't exist anymore. We showed in Toronto at the Funnel. They had a bicycle shop, so there was The Funnel presents at CineCycle. And then we did gigs at schools. In 1983, I went to Europe and I showed at the London Film Co-op. I showed in Berlin at a squat. I took a program of Super 8 films with me from Pittsburgh. That was fun. I moved to New York in '82. I became a regular in the downtown scene in New York. I would go back to Pittsburgh because I was still very attached, but my years in Pittsburgh were only three years. But that was super important. I did two shows in '85 or '86. One was called the Super 8 Motel at The Kitchen, and the other was called Super 8 Solar System. That was at Artists Space. Super 8 Motel was like Beth and Scott B, Lewis Klahr, Joe Gibbons, a whole bunch of downtown Super 8 people. The Artists Space thing I did that with Stephen Gallagher who had been the programmer up at Hallwalls [in Buffalo], and he came down to do the show.

Did you get prints or did you screen camera originals?
I was a camera original person because I used to recut my films all the time—a very bad habit. But I did make prints. I have like two prints of everything from those days. I had some prints made when I was in Europe by Manfred Jelinski. The other prints I made in New York at Lab Link or Kin-O-Lux. But back in the day there was a really amazing

---

"Recently I've been preserving some Super 8 movies and I was trying to look at something on the viewer and I can't even see an image. I mean how did I even do it?! How did I ever know where to make and edit? It's crazy."

INTERVIEW

Dave Markey

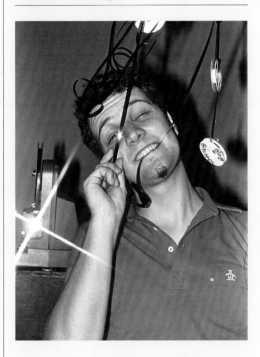

suburban film club and they were into Super 8 anamorphic projection. They would meet in a church basement. We would go out there. Me and some other people would show our movies or they would show theirs. It was like a film club. They always had those Super 8 reduction prints of feature films that they would show. They're kind of amazing. But there was a guy who striped Super 8 film. He would put the mag stripe on it. He had a little machine. I remember getting splicing tape from one of those guys. It was this other world, but I dipped into it and found it fascinating. Just businessmen and regular guys that had this hobby.

What was their response to your work?
They thought we were interesting and weird. It was very different than them, but they liked it. And we would sit and watch *The Ten Commandments* in a reduction print, the last twenty minutes, then we'd look at one of their films. I remember this film and it was the winning film on the home movie anamorphic Super 8 circuit or whatever they called it. It was basically this guy in Malibu who went around his place and showed how fantastic his place was. I remember thinking, "This is bizarre. It's not really a film, but just him showing off possessions and his view." And then we would show one of our films after that. It was eclectic and strange. People that are attracted to film or the arts [are] an interesting array of crazy individuals. There was this woman Tippi Comden. At some point I did a show [at] Light Industry in Brooklyn and she had loaned me a movie she'd made with her students. She worked with deaf kids and she made a Super 8 vampire film with her students. She was part of that suburban group, but she was also part of Pittsburgh Filmmakers. She was a crossover. I still maintain that kind of breadth of influence. I'm not so specifically locked into what is an art movie.

Do you think that's also a Midwest thing? I'm from Detroit and I resonate with a lot of what you're saying. You don't have a ton of artists in a smaller town, so you gravitate to whoever happens to be making things.
That's totally true. And the people in the music scene, we would shoot the bands. Sometimes we'd have film screenings in the clubs. The people in the bands worked in banks or they were waitresses, some of them were just musicians on the weekend. It wasn't like people didn't have aspiration, but it wasn't super professionalized. I'd go to New York and show movies, but I knew that my ability to make movies and what authorized me to make movies was definitely the low-end technologies and this scene I was in and it wasn't a for-profit endeavor or a professional endeavor or a career endeavor. To me it was extremely fragile, this set of relationships that went with the Super 8 and went with the people. The aesthetic that would arise from that was quite delicate to me.

When did you start moving on from Super 8?
My last Super 8 movie is called *Martina's Playhouse*. That's from '89. Around that time, I made *The Deadman* in 16mm. I was trying to make something that was going to be edited and A- B rolled and the processes that allowed you to make prints and get things out. Around that same time, I started getting interested in Pixelvision in the early nineties. And my [Super 8] camera started to get cranky and my viewer broke. My projector, the belt started to slip, and I couldn't get anyone to fix it. So I was just like, "Screw it. It's the end of this." Recently I've been preserving some Super 8 movies and I was trying to look at something on the viewer and I can't even see an image. I mean how did I even do it?! How did I ever know where to make and edit? It's crazy. I used to be able to just hold things up to the light and know where to cut it. Oh boy. ∎

THE 1970S WERE OVER. Punk had given way to hardcore. Major labels, which had dabbled in punk, weren't interested in the nihilistic emissions of this more abrasive brand of rock that was emerging across America. Undaunted by the rejection, a vibrant underground music scene percolated throughout the 1980s. Bands formed and labels sprang up. Touring networks were mapped out. A cadre of zines shed light on the scene. Super 8 was the perfect medium to enter this constellation of cultural endeavors, and no filmmaker better represents this moment and time than Dave Markey. *Desperate Teenage Lovedolls* (1984), was his feature-length Super 8 film starring members of Black Flag and Redd Kross. The film is propelled by teen runaways, fending for themselves, trying to make it in a corrupt

INTERVIEW

music biz. It's a riotous burst of punk rock energy. Markey got the film transferred to video, allowing him to take advantage of the nascent home video market. Super 8's crude, immediate delivery was the perfect vehicle to capture the shambolic nature of the scene that Markey inhabited. As grunge-mania took hold, Markey was in prime position to capitalize. He made music videos for Mudhoney, Firehose, The Muffs, Shonen Knife, and Sonic Youth. In 1991, he was invited by Sonic Youth to follow them on a tour of European festivals where he documented the proceedings. The resulting film, *1991: The Year Punk Broke* is an incredible encapsulation of the American underground music scene moments before it broke wide open, unleashing grunge on the masses. The film features Sonic Youth, as well as the Ramones, Nirvana, and Dinosaur Jr. Like so many folks in the scene, Markey worked across a number of disciplines. Not only was he a filmmaker, but he also played in the bands Sin 34 and Painted Willie, and published the fanzine *We Got Power*.

(L-R): Thurston Moore, Lee Ranaldo, Dave Markey.

When did you start making films?
In summertime 1974. I was eleven. Picked up my dad's Regular 8mm camera. It was a Brownie windup. Figured out how to load that because you could only shoot twenty-five feet then you had to flip the film over.

You were just making films with your friends?
Yeah. I was always pretty good at directing kids before I even had a camera. I had a little crew that would always be doing things around the neighborhood. It would make sense that I would start documenting it. Prior to that I was obsessed with photography through the Instamatic camera format. I came to film through that. I started on the regular Instamatic then graduated to the 110. I remember the transition. Same thing when I graduated to Super 8. I have the memory of upping the format and having more image to play with. My first film, which was a horror film, *The Devil's Exorcist*, was made during the time of *Exorcist* mania, and I was too young to see the movie. The only way I could get to it was to create my own version of it. I thought I would one up the title. I would bring the devil into the title. I used to make haunted houses in my garage so I had things laying around that I could use.

You talk about graduating to Super 8. Did you buy it yourself or was it a gift?
It was a birthday present from my mom. It was my fourteenth birthday. I got a camera and a projector. It would have been a Canon Super 8. Not sound. Graduating to sound was the third step. I had so many cameras. I had a Yashica, I had an Elmo, I had some off-brand Wollensak. I would bring them out on my skateboard. Occasionally they'd get trashed, but they were pretty cheap. People in the neighborhood knew if anyone had a camera they weren't using, they gave me one. At fourteen I had already shot a ton. I used automatic settings at first. I learned more as I went and that was great because I was paying attention and trying to get the best image. With early 8mm stuff, things require a lot of light. If I shot in a low-light situation it wasn't so great, so a lot of things were shot outdoors, in alleyways or on sidewalks. I was doing it myself and learning as I went. A few years in I came across this book, *Guide to Filmmaking* [by Edward Pincus.] It was written by a film instructor and it was about his kids making movies. I didn't know any kids that were making film. I thought, "I'm not alone, there's other kids out there."

How do you go from making *The Devil's Exorcist* with the kids in the neighborhood to becoming part of the LA scene and making the *Slog* movie?
There were a ton of movies in between. *The Movie of Movies* is my *Kentucky Fried Movie*, early *SNL* sketch comedy parody reel, parodying popular commercials, sketches, films, or TV shows. I was fourteen when I did that. At fifteen, I decided I was going to write my first script because a lot of stuff I did pretty much on the spot. When I got to sound I would have dialogue, but we pretty much made up dialogue and usually shot first take. There was no such thing as takes. That aesthetic made it into the first *Lovedolls* film. The first *Lovedolls* film is not too far removed from *The Movie of Movies* or *The Omenous*, which was the first scripted film that I wrote, which was my take on a Mel Brooks-style parody of a genre movie.

You're hitting all the touchstones of growing up in the seventies. Kentucky Fried Movie, Mel Brooks, SNL.
I loved those movies when I was young. I really liked comedy. *Young Frankenstein*,

High Anxiety. Movies I would watch over and over again, and then attempted to make my own versions, with zero budget. From my first film on, I was physically cutting the film. I learned that once I graduated to Super 8 sound, it wasn't so good to have camera cuts when you're trying to have seamless sound because there was always a drop out in the sound. I used a lot of sound-on-sound features. I had sort of perfected the act of editing in-camera. By the time I did the first *Lovedolls* film, there's entire sequences that are like that. The scene on the beach, the "Stairway to Heaven" sequence, with the She Devils coming down and going after Kitty Karryall—all that action was edited in-camera because I knew that I was going to have a continuous song under the track, and I couldn't be chopping it up. I had to anticipate the cuts as I was shooting.

You have to come up with your own edit strategy to get around the limitations of the medium.
I was very aware of that. When I look back, that's the stuff I'm most impressed by. I was able to imagine this stuff on another level, as I was doing it.

Any young kid watching today would have no clue how complex that was.
It was through doing and learning on the spot. And I was documenting my life too and picked up documentarian skills in a similar fashion. I figured out how to shoot things more intimately, making the subjects relaxed enough to actually have them be themselves and not put on an act.

How did you become immersed in that LA punk scene that you end up documenting?
One interest in one band led to another band, which led to the actual scene that was happening at the time. It started out in the seventies getting interested in bands like Devo, the B-52s, Talking Heads, the new wave stuff. That eventually led to hardcore punk. I don't think I would have been able to make that transition from Boston and Kansas without the new wave stuff. Perception is so different now because at the time things sounded so radically different. Now the Ramones sound normal. At the time it was a completely alien culture to what was accepted. I was excited by the more radical things that were going on. I was also a skateboarder and was also coming to it from that. I also really use skateboarding in my filmmaking. A lot of dolly shots from my skateboard. At that time, I was living on top of the skateboard, so it makes sense.

Talk about the *Slog* movie.
I wanted live recordings of my favorite bands and not a lot of people brought Super 8 cameras to these shows. No one did. I was doing it for my own gratification. I really liked these bands and I wanted to see them again and again, whether it be The Circle Jerks or Redd Kross or Wasted Youth or TSOL or any of these bands that I was getting into in '80–'81. I was starting to grow up. I moved out of my mom's apartment. I was in a band that was active. I had done a fanzine. I had ten years of filmmaking under my belt prior to *Desperate Teenage Lovedolls*. Maybe not all of it was good or watchable or maybe not all of it was anything that I showed to more than a few people but…I was nineteen years old when I started making that film, my last year of my teenage years and very aware of it. You know, post high school trying to figure out what the hell I'm going to do with my life. All that comes into play in terms of the content. Things are a bit darker.

At that point, since you're doing zines and you're in bands, is it a stretch for you to get that cast into your film?
Through the fanzine and through being in a band and playing shows with Redd Kross and Black Flag, you end up making friends with band members. It naturally happens where like, "Oh yeah, I'd like to be in your movie." Or maybe I ask someone to be in the movie and it goes from there. It was thrilling because I idolized Dez Cadena. He was singing in Black Flag at the time. I just thought he was the fucking coolest and was so excited that he was going to make himself available to be in my movie. It made me believe a bit more in what I was doing.

Did you have a plan for showing that film?
I was going to show it like how I showed my previous films on Super 8. *The Omenous* was feature length and it had three reel changes. I had two Super 8 projectors. I used 200 foot reels and I did it like how it was done in big movie houses where projector A is running and I knew at what point in time I was going to need to start projector B. That's how we first screened *Lovedolls* at the Lhasa Club in Los Angeles. They had a small theater that sat about 80. It actually had a projection booth.

Do you remember what projectors?
The Elmo S1 180E two track projector. This one you could record on the outer track as well. I put sound effects there, sometimes music cues.

You would do sound-on-sound on the main stripe and then also do another batch of sound on the balance?
I did that with the *Lovedolls* film.

The first *Lovedolls* movie gets out in the world. Is it just from you touring with the film?
It's the first film that I made that got any critical response because it was the first film that was actually presentable to screen. The *Slog* movie was, but outside of showing it on the wall at Oki-Dog because there was a power strip back there (I brought my

INTERVIEW

projector out there one night and that was the performance). Besides a couple of punk rock parties where I screened the film, it was the first presentable film that I made that people outside of my small circle of friends could actually see. Before the film was finished, I screened for a few people that were involved with the LA Weekly, namely a writer/musician Craig Lee who was in the The Bags. He was enthusiastic about it and wrote about it. I was living in a recording studio at the time called Spin Head and I would set the projector up in the control room, it had a glass window, acted as a de facto projection booth. I would invite people and a lot of writers would show up. It just seemed to happen. A positive critical response. It took a year or so to build. In July 1984, the film had its first screening at the Lhasa Club and it was sold out because of the fact that there was press in the LA Weekly and the LA Reader. Don Howland wrote the most significant piece of press for it for The Village Voice. There was critical attention, smalltime, coming from the weeklies on each coast. But it was enough to have the telephone ringing. I really beat the streets and did screenings in clubs because I was already used to doing that from booking shows for my band. Also there was a video theater in West Hollywood called Easy TV and they had their own 3/4" U-Matic edit system there. I did all the post for the sequel, Lovedolls Superstar, there. I did a ton of screenings of the first Lovedolls there and brought other Super 8 films in there. I brought Richard Kern's films into that place. We screened them on double bills with my films. We always sold the screenings out, between 50 and 100 people and momentum built. Punk fanzines played a crucial role in spreading the word nationally. Also, I did a soundtrack album. It took almost a year to come out after the film, but the album was responsible for spreading a lot of word. I have friends that lived in various parts of the country and told me stories of finding it in a video store in like upstate New York. You know, kids stumbling in and seeing the videotape and just being attracted to the graphics on it—finding the film that way.

When did you put it onto video?
Right away I put it on VHS and Beta. I sold the tapes through consignment at record stores or at shows or through our fanzine. We'd take out ads in our fanzine or in other fanzines. Eventually sold a few hundred copies. It did get the film out there and it was the kind of film that people would copy for their friends, so one copy would get to twenty-five people. There was a lot of that going on in underground circles.

Where did you get the film transferred to video?
Sunset Sound on Sunset Boulevard. They didn't have the best system, but they were the only ones that I had heard about. They had a rank, pull-chain transfer, the [projector] pointed directly into a video camera and it went to U-Matic 3/4". That was dubbed many times. I would do rounds of twenty-five at a time. They were fifteen dollars a piece, and I'm selling them for twenty bucks. By the time you tack on the cost for mailing it was a break even proposition. A year in I was approached by a couple of different distributors. I ended up going with a company called Hollywood Home Video. They ended up moving a couple of thousand copies of the movie. They got distribution so it wound up in the cult section of Blockbuster Video. Simultaneous to all of this I was already in production with Lovedolls Superstar. People were coming forward wanting to lend a hand and be a part of it. I used what little moneys that were made for the budget of the sequel. Compared to the first film, which cost a few hundred dollars, I now had about ten thousand dollars to play with, and that's how I got my sequel made.

Do you think Super 8 lent something to those movies or, looking back, would you have rather shot on 16?
Again, I was just using what was available, but the aesthetic of Super 8 ended up working out fantastically. That gives it its own feel and aesthetic. It might not have worked in 16 or 35. Maybe the freedom that was involved with it, where you're free to do whatever it is you want. You're still working with film, it's just a smaller version of the real thing.

When you get to The Year Punk Broke, is all of that Super 8? Was the label excited about that plan?
It's all Super 8. The label was really out of the picture. Sonic Youth and I put up the meager funds for it. It was a very last-minute decision to bring me along. Thurston initiated it and I didn't have a whole heck of a lot of time to prep. It was just a couple of weeks.

You're there as a crew of one?
Yes. I had purchased a Beaulieu 7008, which was the state of the art Super 8 camera. I used the Angenieux 12-120 lens that I purchased through The Recycler. It gave me a great look. I did a couple of music videos for Sonic Youth for the Goo album that made The Year Punk Broke possible because they turned out so great. I also did a music video for Firehose just prior to that, and one for Shonen Knife with this new camera and it was like, "Oh wow, Super 8 is starting to look really good now." It sort of caught up to 16. Up to that point there were very few Super 8 cameras you could swap lenses out. Most lenses were fixed, so that was a revelation. Also, the camera allowed for the 200-foot mag so I could shoot more than a few minutes at a time, which was crucial when documenting live music performances. It was a short tour, only two weeks. It's crazy to think that I got all of that in under two weeks, just myself and a suitcase of Super 8 film.

Was it single system?
Yes. A microphone that went over the top of the camera. The total budget of the film was in the neighborhood of thirty-five, forty thousand dollars. For the budget of a low budget music video I produced a feature film. Of course it all started with the Super 8. I bought about nineteen hundred dollars' worth of Super 8 film that filled up the suitcase. I bought basically eight, nine hours' worth of Super 8.

What's the thing that Super 8 really afforded you to do as a filmmaker?
The absolute freedom and accessibility of shooting it. The unpretentiousness of it. The fact that it was something really easy to do for not a lot of money. From the seventies through the nineties I was basically dealing with the same pricing of film and processing. Technology moved slower back then. Before computers became involved with the filmmaking process, any kid could pick up a camera and create something. It's radical to me. The look. The feel of it. The sound of the camera running while you're filming picking up on the soundtrack. All of these little things that end up tied to the overall aesthetic of much of the body of my work, all those little things, I love all of them. They all make me feel warm and fuzzy about the format. I don't know if I would have been a filmmaker otherwise. I don't know how I could have come to it. Maybe I would have. I only know the life that I lived and the way it unfolded for me. But I just remember the excitement that I felt. I shot a film, then I had to go put it in for processing at Fotomat, and it took a week for the film to come back. I remember counting the days down until I would get that 50-foot reel back. Then I would anxiously get to that Fotomat kiosk on my skateboard and before I'm even home, I'm already holding the film up to the sun looking at it, looking at the image as I'm on my skateboard going down the street. I have total fond memories of that and it's all very precious for me. ■

INTERVIEW

G. B. Jones and Bruce LaBruce

G. B. JONES AND BRUCE LABRUCE gravitated to the Toronto arts scene in the late 1970s and early 1980s. Jones was enrolled in the film program at the Ontario College of Art & Design, while LaBruce studied at York University. Like John Porter before them, they were buoyed by the experimental film scene at The Funnel. However, both Jones and LaBruce were interested in more than just filmmaking; they were making zines and playing music. Jones was a founding member of the seminal queercore band Fifth Column. Jones and LaBruce lived together in the low-rent, east end of town, and their house became a hive of DIY activity, a fertile ground for all their creative endeavors. When they started making films, Jones and LaBruce worked on each other's projects and starred in each other's films. Their films lived on the edge, contained a punk snarl, and were unapologetically queer. This mix made screenings a challenge. In response they started their own screening series, the *J.D.s*, which featured their work along with that of like-minded coconspirators.

What got you interested in film?
Bruce LaBruce: I had parents who were cinephiles. They took us to movies constantly. They were rural farmers but they knew everything about classic Hollywood. They instilled this interest in me in films. I was going to be a critic from a very young age. When I left the farm, I took production for two years at York University. York is known for its activism, international studies, queer studies, and social awareness.

G. B. Jones: The first place I remember discovering underground film and art was

Bruce LaBruce on the set of Super 8 1/2.

at the library. They had the magazine *Films and Filming* that came out in the seventies, or maybe *Film Culture*. I can't remember, but it was a magazine devoted to film as an art form. It covered European filmmakers like Fellini, Pasolini, Bergman, Godard, Varda. It also included experimental filmmakers like Warhol, Jack Smith, the Kuchar Brothers. I became interested. It wasn't just that they were making films, but they were leading a life that was resisting the norms of the day. That was exciting to me. Rebellious teenager.

Where did you grow up as a teenager?
GB: North York and Martin County. Those places are close to Toronto, so I could take buses and subways and get to Toronto. But when I did get to Toronto, there was no underground scene here. It would be quite a few years later. Then I came downtown to go to art college. One of my instructors was Ross McLaren. He did a lot of Super 8 film in the seventies and eighties. He had done

INTERVIEW

Crash and Burn, that was about the punk thing at that time in Toronto. I saw that film and loved it. He was teaching at OCAD and was a member of The Funnel. I started going to The Funnel every weekend. That would be 1978, '79. I borrowed my brother's Super 8 camera and started making Super 8 films because it was such an easy way to make films. The first film I made I went to stay at a friend's. A bunch of friends got together, and stayed over for a week at their apartment. I started filming all these silly things we did, features of their apartments that were new and novel to me. You could watch on the television and see everyone who came to the door. I thought, "Oooh, I'm filming that." Different things about modern living. It all was informed by J. G. Ballard's *High Rise*.

BLB: In the eighties, [York] was a hotbed of political activism, and feminism, and studies about racial struggle. I took two years of production because I wanted to know how films are made, even though I intended on being a critic. First year we did photography and second year was Super 8. I made some extremely bad films as a student. Although there was one I made with a couple of female friends of mine, a crazy experimental film. It was this female friend floating down a river on a raft. It was about her and another guy seducing a young girl, and we would intercut the scenes with shots of bees and insects. After two years I went into film studies. I was lucky to have, as my mentor, the great film critic Robin Wood—gay, Marxist, feminist film critic. Him and a few professors and students started a radical film magazine called *Cine-Action!*

GB: I started doing a kind of documentary on stuff that was happening where I

Filming The Troublemakers *(L-R): Joe St. Pierre, G. B. Jones, Bruce LaBruce.*

had grown up in Martin County. I have a friend who lived on a farm and knew this creepy guy who was driving around in dune buggies, and would meet girls hitchhiking and bring them home. He had twenty people living in his house, and there were tons of drugs, and he ended up shooting somebody. Her personal experience with him was horrible. I started to make a documentary and she took me to the house. It was empty because there had been a trial, and everyone had to leave because he ended up going to jail. I filmed the house and the grounds. It was like a docu-drama, because I had fictional elements in it as well. I started [*Unionville*] in

'79, and didn't finish until '83. That was the first film I showed at The Funnel. In those days you could get Super 8 sound, but I decided to do a voiceover. I put all the sound on cassette. Of course that got wrecked over the years so I had to redo the sound.

BLB: I hung out at the punk scene downtown and that's where I met G. B. Jones, Candy [Pauker] and other members of Fifth Column. We all worked at an all-night dessert restaurant called Just Desserts. It was one of the only places that would hire crazy junkies and punks and artists and musicians. The owner would let us be ourselves and dress how we wanted. It was a kind of place where you would go to get abused by the staff. Everyone had their fanzine, GB and I had J.D.s, Candy had a fanzine called *Dr. Smith*, and GB and Caroline Azar did *Hide*. And we're all working in Super 8. I made my first serious film that I wasn't embarrassed about in '86 and then started making more Super 8 films between then and 1991, when I made my first feature length film, *No Skin Off My Ass*, which was completely shot and edited on Super 8.

Can you talk about some of those early films?
BLB: I did *Boy, Girl*, which GB is in, and *I Know What It's Like to Be Dead*. Then Candy and I did a collaborative one called *Bruce and Pepper Wayne Gacy's Home Movies*. Then I made one called *Slam* which was shot in a mosh pit, intercut with found porn, to a Carpenter's soundtrack. I appeared in Candy's *Interview with a Zombie*, which was a gay zombie film.

GB: Fifth Column started in 1980, it wasn't officially called Fifth Column. Before that we'd been a band called Second Unit. Of course you know what a second unit is, it's the people on a film crew who go out and film the location. Anything to do that needs to be filmed that doesn't include stars. We were the second unit. When Caroline Azar joined, we changed the name to Fifth Column in '81. We were playing benefits at The Funnel and John Porter (see John Porter Interview, page 145) suggested to us to do Super 8 film loops while we played. John was making a lot of his own Super 8 films, and we were going to see them. We would try to wear clothing that would be good for having [John's] films projected on us. We would often wear white or put a white sheet behind us. We had films for at least twenty different songs. We'd get together and we'd talk about what we wanted on film. Caroline was involved heavily because the films would have to reflect her lyrics. John shot all the content.

It seems like there's a scene developing in Toronto in the early to mid-1980s.
GB: Definitely. I started filming *The Troublemakers* in 1985. It was me Caroline, Bruce, Candy, and Anita from the band. We were all living in a ramshackle, rundown house. I started thinking about Super 8, and the associations it had for people. I became interested in the idea that most people used Super 8 to make home movies. I thought it would be interesting if I made a home movie, except that my concept of home and family was so radically different. I wanted to experiment with associations that Super 8 had and challenge that idea of what it was supposed to be for and about. It's the same time Bruce was starting to make his films and I was in four or five. It was a very exciting period in our house. The band was practicing there, Bruce was making films, I was making films, Caroline was working with John on the Fifth Column films, and Candy was taking photographs and making films. We also started to make fanzines as well. We were the happening studio.

BLB: We were just documenting ourselves and each other with Super 8, doing what we do every day—giving each other piercings, shaving our heads, eating dog food out of bowls, lol. It was pre-internet and pre-digital, so we were well ahead of the curve in terms of this kind of self-documentation.

How were the films and the zines and the J.D.s screenings intermingling?
BLB: They all came around the same time, and it was an aesthetic that came out of the fanzines. My Super 8 films were very much based on a collage sensibility. I would take found footage, a well-known experimental technique, but take found footage of porn that I would find in a bin in a used bookstore, or footage I'd take off the TV, or people would just give you random stock footage, and I would intercut that into the footage that I took and construct a narrative out of it. It was the homocore, queercore aesthetic and politics shared by the fanzines and the films. We would promote the films in the fanzines. I didn't market my early short films so much, but we did have screenings. There were various places in Toronto like art galleries, the Purple Institute, The Funnel, and bars which would show our films. Candy and GB were doing a lot of détournement. They would take comics and queer them. GB used Archie Comics, and through editing and rearranging them, she'd turned them into queer narratives between Archie and Jughead.

Were you getting your other films out into the world or were you having trouble because of content?
GB: That's one of the reasons we started doing the J.D.s Film Night. We did them in Montreal, Toronto, San Francisco, New York. We knew a number of people making Super 8 films and started showing them. I was showing *The Troublemakers* as part of that, Bruce is showing films like *No Guff, Egri, Bruce and Pepper Wayne Gacy's Home Movies, Boy, Girl*, which is named after one of the Fifth Column songs, as well as [films] by Suzie Richter and David Gravelle. The Funnel had closed by that point, so I didn't have access to show my films there, but because we had a readymade program, film festivals would pick us up and show the whole program. When we went to Montreal it was part of the Montreal Gay and Lesbian Film Festival. In San Francisco we just got in touch with people at a bar, and held it at a bar.

Do you remember what bar?
GB: It was the Crystal Pistol.

That place is now a super fancy restaurant. Every time I walk by, I think, "People have no idea what went on in this place." If you could peel back the layers, you wouldn't want to be eating a fancy dinner there.
GB: But that's the kind of places that we went to, because they were amenable to doing stuff. Caroline, at the bar she worked out of Toronto, also had a Super 8 film night. Her bar was called "Together." It was a lesbian bar. But these were the kind of bars we ended up in because they would let us come in and do crazy film screenings.

In the mid-1980s there was a real resistance at the film festival level to screen things with a punkier edge. Were you finding that as well?
GB: Oh yeah. The gay and lesbian film festivals in Toronto were not interested at all. None of them would show our films. Our films were never shown at a gay and lesbian film festival in Toronto, until maybe the 2000s, at least twenty years after they were made. We had so many different problems. At that point, all our mail was being opened by the Royal Canadian Mounted Police, and our phone was bugged. When we showed our film in Toronto, we showed them at this anarchist space called the Purple Institution. The first time that *The Yo-Yo Gang* showed

INTERVIEW

was at Hallwalls in Buffalo, which was an amazing space devoted to experimental film and performance art. We went to the opening night and there were Christian protesters, protesting the screening. We had to walk through a picket line of Christians, all upset that there was a queer film night happening. This is the kind of thing that was normal then. We had to be careful about what labs we took our Super 8 film to get processed. I'd always have to go, and talk to the people. I had to say, "I have this film, it's got weird stuff that you might think, 'What's going on?' I have to reassure you I'm not making porno movies. It's just people having fun and getting drunk." I don't know how many times they looked at me strangely and I'd be like, "Okay, I better not leave that at this lab." They'd phone people up, report it, and then seize the film. We had to be careful about mailing our films because of course they might never reach the destination or come back. I don't think I ever mailed my films anywhere. When we had the screening in London, England, somebody from Toronto took them there. When we went to San Francisco and other places in the States, we were traveling with our films.

These experiences informed your work too, right?
GB: Of course, because you don't want to be censored and you don't want to be repressed by these cultural forces. If you can work at home, do your editing, and then set up your own screening, you can circumvent the entire censorship process. One of the reasons The Funnel had so many problems was that they were fighting an ongoing battle with The Censorship Board of Ontario. They refused to submit any of the films to the Censorship Board, to be cleared or not. They said, "We're not doing that," and the Censorship Board harassed them endlessly for the ten years they were open.

BLB: When I was doing the transfer of *No Skin Off My Ass* from Super 8 to 16, it was at a lab in Toronto. I had a good relationship with him personally, but when he saw the film that I was trying to transfer he called the cops. Technically at that time, I don't know if it was a law or just a policy, but they would call the cops if they saw any sexually explicit material. This would be '90 or '91. He told me the cops had been there and they wanted me to cut three specific scenes out of the negative. That was nudity with violence, bondage, and sucking of toes. I was totally freaked out and I didn't want to cut it. But because I had a relationship with this guy, he basically said, "Look, by law I had to call the cops, but if I leave the film on the desk here and have to run into the back to attend to something…you know, anything could happen." So he left and I grabbed the film.

***No Skin Off My Ass* is a feature—was that a big jump for you?**
BLB: For *No Skin Off My Ass*, I was just doing everything myself. It was the same thing I did with my short films. I shot it on Super 8 and edited the original. I would transfer it myself onto cassette. I made the VHS cover myself at a printing store, and then would put ads in the fanzines and get the orders and mail them myself. It was do-it-yourself from beginning to end. Jürgen Brüning was the visiting curator at Hallwalls Gallery and he would come out to Toronto to scout for work, and that's where he saw our films. He became my producer. I asked him to blow it up to 16. That was the big jump, because then I was able to submit it to film festivals. At that time, which is like 1990, '91, gay and lesbian film festivals were exploding all over. Suddenly every small city would have a festival. So it got around and became a cult film. And then I shot my first film on 16mm, which was my following film, *Super 8 1/2*.

Is there any sync sound in *No Skin Off My Ass*?
BLB: No, it's totally dubbed, and badly. Really badly. In that film I had a rough script, but there was also improvisation. I used to show it in bars on Super 8 projectors with the soundtrack on cassette. Every time I started it, I'd try to start it at the same place, but it would always be slightly different. It was a different viewing experience every time. When I finally dubbed it, we did go into a recording place, but we didn't even bother to try to make it sync.

When you were making those early films was the notion of working in 16 on your radar?
BLB: I dropped out of production after my second year because I was a total technophobe. I didn't think I'd ever be able to figure out how to make a feature film. And it was expensive. For me, Super 8 was so easy technically and it was relatively cheap, so it seemed to be the only option.

At some point you write a manifesto that was in *Maximumrocknroll*.
BLB: GB and I wrote it. Maybe '89 or '90. It was called *Don't Be Gay or How to Stop Worrying and Fuck Punk Up the Ass*. We were big on manifestos. It was basically describing what we were doing, what we were up to with queercore. Our whole thesis was about how the early roots of punk were open to sexual diversity, not only queer but transgender. We'd be citing Southern California punk bands, digging into musicians that had queer references in them like the Ramones and Patti Smith, and homoeroticizing musicians like Iggy Pop or David Cassidy. It was our diatribe against the punk scene, which had been becoming increasingly homophobic, partly because of hardcore music and fashion. The pit became very aggro and macho, and homophobia was prevalent. It was our mission statement really.

INTERVIEW

GB: That was our manifesto introducing the world to queercore, which was a merging of queer culture and punk culture. For me that was familiar territory. I had been interested in experimental film, and as you know, so many of the people involved in experimental film were queer or gay. Jack Smith, the Kuchars, Warhol, and Derek Jarman. In the experimental film world, it was no big deal if people were queer. It was only in the larger world that it was incredibly important to people. So we had this idea about queercore. Rather than try to isolate the different worlds we belonged to like the music world, the experimental film world, or our regular jobs, we thought it would be better to integrate it all. But it led us into a confrontation with the sensibility of the day. They weren't really ready to accept queer people. So we just got into our little movement, queercore. The Super 8 films were a part of that because I wanted to make sure that people were able to create things, whether it was a band, or a fanzine, or films—that it was easy for them to do and not financially impossible. Super 8 was something I was promoting as a vehicle to create your own film, to be able to show them to people, and create your own culture. I thought punk was about creating your own culture rather than buying products that were presented to you on TV and in magazines. It was about rejecting that and creating your own culture. I thought that queer people needed to start doing that.

Talk a little bit about *The Yo-Yo Gang* and *Lollipop Generation*?
GB: With *The Yo-Yo Gang* there was a magazine store across the street from us. I'd look at old magazines and noticed that so many of them had ads for like these little Super 8 movies. They usually starred Bettie Page, and she'd be dancing around or she'd be spanking somebody. I thought this is a fascinating history of the use of Super 8 film.

With *The Yo-Yo Gang*, I decided to make a movie about girls in a gang, but at the same time use what's considered exploitation fetish films as a template. So, in *The Yo-Yo Gang*, every scene revolves around a different fetish, such as girls fighting, which was a popular theme in those films, or a woman smoking, or bondage, or spanking or piercing, tattooing. Anything that I thought would appeal to someone with a fetishistic interest, I put in the film. But at the same time, the overall message was one of emancipation. I was trying to set up a kind of friction in the film between these two impulses—emancipation and fetishization—that I thought would be interesting to work with.

Lollipop Generation I started filming on the many tours Fifth Column took. Another use of Super 8 was people loved to film their trips. I thought, "Oh, I'll make a travel film." I started filming all the towns we'd go through. It was only after I thought I'm going to have to try and put a narrative on top of this to give people a reason to watch it, because in real life people don't want to see your Super 8 travel films. You can trap a certain amount of friends in your suburban basement and force them to watch that kind of thing, or you could in the fifties. By the time the nineties came along…not so much. So I had to think of a little narrative. I thought it would be an interesting premise if the people in the film decided they weren't going to be exploited by this evil pornographer in the film, but would make their own films.

What part of the process did you like the most?
GB: I like everything about it. I liked being on film sets and I would be helping people hold lights. I did people's makeup. I did their hair. I ran and got clothes for them to wear. Whatever people wanted or needed help with doing, I was always excited to help. I found it exciting and invigorating. It was like life but amped up ten times. ∎

Martha Colburn

MARTHA COLBURN IS a force of nature, and her films back that up. They are cut and paste, slash and burn animated affairs that pulsate with manic energy. Hints of Lewis Klahr, Larry Jordan, and Terry Gilliam can be seen, but Colburn's aesthetic is wilder and punkier, thrumming with propulsive lo-fi music. She started making films in Baltimore and burst onto the underground, experimental film scene in the 1990s. Colburn currently resides in Amsterdam, where she is producing increasingly ambitious digital animations with her trademark stop-motion style.

What drew you to Super 8?
I chose Super 8 because it was the highest quality image at the cheapest price. Actually, someone gave me my first Super 8

INTERVIEW

camera, so it was free. Video looked bad and 16mm was too expensive. I came to love the color and quality. If I wanted to play with 16mm or 35mm, I would hand paint it and refilm it, frame-by-frame, to Super 8mm (for viewing) by taping a magnifying glass to my Super 8 camera lens. I don't let things stand in my way. To make a good film, I knew Super 8 has a small amount of grain, so I always shot very close-up.

What did your setup look like?
My dad's tripod, from when he was an entomologist in university and had to photograph bugs, and two clamp lights taped to chairs. I filmed exclusively on the floor.

What camera did you use, and did you have an animation stand?
Canon 1014 XL-S. I never had an animation stand. I invent my own way of doing things. Now I just find some glass on the street and prop it up on some books or CDs. A stand to me looks like a torture device, like in those torture museum advertisements in Amsterdam. I want to be free of any "traditional" trappings of being an animator. Super 8 fits this attitude. It is amateur and without any expectations.

Can you talk about the process of some of those early films? How long did it take to shoot a film?
I was making around ten films a year, so one or so a month. And then I made longer films and it took longer. That's a simple formula. I would make a series of collages and paintings, using oil paint and magazines, on the backs of book or old record sleeves. Then find a song—usually being made in the back of my warehouse in the recording studio of Megaphone Records. Sometimes I would make music too. Then I'd make a film.

How did you manage to make so much in such a short time?
I worked construction sites as a painter, and also banks, steakhouses, taco trucks—you name it. Even the US Capitol. Then I had time to work.

Music was such a huge part of your films. I assume all the films were animated in-camera and then the sound put down in post?
Yes! From quarter-inch tape. The music was mixed to cassette and then transferred from my cassette deck to the mag stripe on the Super 8.

Did you shoot sound stocks? Or post stripe?
Sound stocks only.

How were you adding the sound in post? Projector? Viewer?
My projector.

Were you concerned about frame accuracy?
Oh yeah, I count them—making the films and after. That's what separates the normal filmmaker and the animator.

Who was inspiring your work at the time?
John Waters, tENTATIVELY a cONVENIENCE, Skizz Cyzyk, these were all these Baltimore and DC filmmakers and musicians. We were making music, musicians were making films, I helped run a cabaret, everyone is doing stuff. But I discovered the whole history of film through the library collection of 35,000 titles on 16mm.

Your films showed around a lot. Were you getting prints, or just showing originals?
All originals. All fifty splices per two-minute film. All Super 8. I even did a whole USA tour from SF to Baltimore with this master reel. I hired Brodsky and Treadway to make Beta SP transfers of my main films, before I went on tour with the originals. Then several of them went on tour and were later transferred. But even now, I still have Super 8 films, which are some of my favorites, which have only screened one or two times in basements in Baltimore. I still have transfers to do. I love the originals though.

How did Super 8 allow you to develop as an artist?
Super 8mm was the first format film I ever shot. I learned about film through using found footage 16mm, and when I was able to create my own film with Super 8 it was so inspiring. I did not think of Super 8 as an amateur format, because projected it is powerful. I wanted to make the most fully realized film I could—in sound design, color, editing, and artwork, and contain it all in this minuscule format. ∎

"I wanted to make the most fully realized film I could—in sound design, color, editing, and artwork, and contain it all in this minuscule format."

Top (L–R): Cats Amore *(2000),* What's On? *(1997). Bottom (L–R):* Spiders in Love: An Arachnogasmic Musical *(1999),* Cats Amore *(2000).*

INTERVIEW

Matthias Müller

MATTHIAS MÜLLER IS a German experimental filmmaker with a rebellious attitude. He started making films in the early 1980s and embraced Super 8 for its outsider status. He helped form the Alte Kinder Film Collective in 1985, which showcased and distributed the work of a new wave of younger German filmmakers who were operating in a DIY and post-punk vernacular. They eschewed the old-boy network that had built up around the established experimental film scene and were content to forego museum shows in favor of alternative spaces like squats and bars. Müller has produced an impressive body of work, often working in the found footage vein. He also hand processes much of his film, embracing the imperfections that can arise from this technique. He currently teaches experimental film at the Academy of Media Arts (KHM) in Cologne.

What drew you to film as a way to express yourself?
Starting to work in film was an extension of my previous artistic interests and practices such as drawing and paper collage. As a time-based art, it allowed me to include a crucial new element to my work. Moreover, it offered the opportunity of opening up to artistic collaborations. I had watched a couple of Andy Warhol films as a teenager; I guess it was this inspiring experience that made me grab the film camera.

Why did you begin making films in Super 8, as opposed to 16mm or video?
I was nineteen when I made my first film, and I had not had any education in film up until then. A simple and inexpensive format such as Super 8 seemed perfect for entering and exploring this new field: it was learning by doing. There was a generational shift in experimental film in the early eighties; 16mm and the fetish of the Bolex were mostly utilized by the older generation who were, by that point, striving hard for canonization. This old boys' club, stuck in its formulaic routines, seemed like a gated community to us. Why try to throw open the doors to a place where you do not want to be? At the same time, video art radiated its own cool distance and was, by then, gradually being accepted by the art world; it was becoming more and more institutionalized. We were striving for a more intense, immediate experience. Working in a marginalized, disparaged and substandard medium such as Super 8 implied the promise of working outside hierarchic power structures, of defining our own place.

Were you seeing a lot of work being produced on Super 8? Do any films or filmmakers stick out as being an influence on your work?
Early on, I started attending film festivals—and I did so excessively. While it took the larger ones a long time to include Super 8 in their programs, there were countless smaller ones sprouting up in the early eighties, and quite a few of them were exclusively devoted to Super 8. Me and my friend (and first collaborator) Christiane Heuwinkel were regulars at such events: watching new work, making new contacts and helping to expand the Super 8 network. The films that had the biggest impact on us, such as those of Die Tödliche Doris, Uli Versum, and Schmelzdahin, we then programmed for the local museum of modern art: Kunsthalle Bielefeld.

In the essay you wrote for the book *Derek Jarman Super 8*, you talk about "Super 8 as a medium that allowed you to live according to ascetic principles." Could you expand on that?
In his 1984 book *Dancing Ledge*, Jarman described my Bielefeld neighbor, Padeluun, as a man "who believes artists should work, take simple jobs, receive no funds from state or individual beyond what is necessary for the simplest existence." Returning from London, where he had met Jarman, Padeluun brought back the idea of Super 8 filmmaking. Since we strived for total artistic freedom and independence, we tried to keep the biggest distance possible between us and the industry and official institutions. Our films were mostly self-financed, and their production did not come with the promise of the red carpet or financial gain. Thus, there was a large discontinuity between our growing reputation and the fact that the work didn't come close to paying the rent. However, we had deliberately chosen this kind of "poor cinema" and were passionate enough about it to want to share the experience with as many others as possible. This is why we founded a small distribution company, Alte Kinder, in 1985 and why it was based on the principle of at least one filmmaker

attending and introducing the screening. We focused on our own films first, and later included works by other filmmakers, such as Schmelzdahin, Jeanne Liotta, or Owen O'Toole. This work was unprofitable, but the financial backing of regional film-funding bodies and institutions such as the Goethe Institute helped us expose our films to audiences worldwide. All sorts of unconventional spaces were turned into temporary venues for our films: squatted houses, bars, even antinuclear camps. The experience was thrilling, and the responses were rewarding, but finding pleasure in this depended on the acceptance of a low-key, somewhat "ascetic" lifestyle.

What did Super 8 offer you that other formats couldn't?
Super 8 was booming when I started out and the community was growing quickly. Being part of such a movement meant many new contacts, global exchanges and a growing number of alternative venues; it opened up fresh worlds to me. No other artistic medium seemed to offer a membership of such a large cross-cultural and noncompetitive network. Due to its unprofessional nature, Super 8 was one area within a larger DIY culture, influenced by punk and post-punk; its multiple intersections with other kinds of artistic practices turned out to be incredibly fruitful. One critic called it a "creative conflagration" and it definitely was a blast! The choice of Super 8 alone was a statement in itself. This gauge was the key to an experimental, risky approach to filmmaking, a clear commitment to counterculture. The expenses were low and making mistakes did not stop you from continuing. On the contrary, faults and failures often rendered the most exciting or promising aesthetic results. The painterly, pulsating, grainy textures offered unexperienced viewing pleasures; the film stock's sensitive, vulnerable

Top (L-R): The Memo Book (1989), Epilog, by Christiane Heuwinkel & Matthias Müller (1987), Pensão Globo (1997). Bottom: Pensão Globo (1997).

surface made us discover and celebrate perfection in imperfection. My attempts at hand-developing were amateurish and clumsy, but it was this rough-and-ready quality of the footage that lead to the exact visual appearance I desired in the end. Throughout the eighties, I explored various techniques of reframing and refilming. This interest culminated in Pensão Globo (1997), a film that fuses two perspectives into one and the same motif in one constantly refilmed double-projection. Over the course of the eighties my films became more controlled and refined, until in 1989 I made my first openly autobiographic film, Aus der Ferne—The Memo Book, which opened the door to a completely new period of my work. Stemming from a life crisis after a former lover died of AIDS, the film follows the

INTERVIEW

stages I went through in dealing with this trauma. Conceived as an intimate, visceral long-term project, only Super 8 could offer the freedom and flexibility it required.

Were there limitations to working with the medium that proved frustrating?
Indeed. I started working with musician Dirk Schaefer who had already composed soundtracks for my films in 1984 and who accompanied my productions for almost three decades. To me, sound had meant a crucial component from the very beginning, but no matter how immaculate the recording, the moment it got transferred to that narrow-gauge magnetic strip the quality degraded drastically. What's more, Super 8 can only be screened appropriately in smaller venues—unless you have an extra-bright projector on hand. However, I did decide to continue working with Super 8 throughout the nineties, but had my films blown up to 16mm. This was from *Aus der Ferne—The Memo Book* (1989) onward, with the soundtracks directly being transferred to the bigger gauge. *Sleepy Haven* (1993) was entirely shot on Super 8 and hand-developed, but its footage needed to be blown up and edited on 16mm. The more elaborate my work became, the more it needed to be presented in spatial situations that did justice to the quality, which made me shift from improvised projection spaces to cinemas, museums, and galleries.

Did you work with Super 8 found footage and collage?
Working in a medium considered obsolete and employing found footage have one thing in common: it is looking back at things. To me, appropriating my dad's camera and swiping other makers' footage went hand in hand from the very outset. You can even find bits and pieces of found footage in my very first film made in 1979. I do not think I ever included Super 8 found footage, but there was a lot of refilming stuff from the TV screen. Those finds were often of a random nature—like my find of the opening shots of Michael Powell's *Peeping Tom* that I saved from the dust of some Roman piazza after an open-air screening and integrated into my film *Final Cut* (1986). Nowadays, access to even the most remote and obscure footage depends on only a couple of mouse-clicks, so my use of found footage is based on elaborate and extensive research. For one of our joint projects, Christoph Girardet and I researched more than 500 feature films. In the pre-digital age, getting hold of specific footage was much more difficult, which is why I mostly used what was close to hand. In those days I would, for example, lie on a bed refilming a TV broadcast through my legs without knowing that particular movie or program.

Can you talk about the German Super 8 collective Schmelzdahin?
"*Stadt In Flammen* is one of the most volcanic films I've ever seen," my filmmaker friend Owen O'Toole wrote in a review. Luckily, I was able to attend the premiere of this seminal film back in 1984; its screening made me want to meet the makers instantly. We became friends. In their collaborative work, the three members of Schmelzdahin explored the possibilities of the bacteriological decomposition of film footage; however, they also treated their material mechanically, exposing it to overheated projector bulbs or shredding it to dust in a self-made machine. The results were stunning, and they contain a perfectly timeless beauty. Carrying on the experimental tradition of physical film, Schmelzdahin quickly chose Super 8 as their favorite gauge. Working together for almost a decade, they also developed an amazing live film performance: *Wir lagerten wie gewöhnlich um's Feuer*, that allowed the audience to simultaneously witness the artists treat a looped film stripe with all sorts of chemicals and watch the effects of this on a screen. Schmelzdahin's body of work is a tribute to the magic of alchemic transformation, the beauty of impermanence. We were happy to have three of their films in distribution—*Stadt in Flammen* (1984), *Weltenempfänger* (1984), and *Der General* (1987)—and to artistically collaborate with them on a project initiated by Owen O'Toole: *The Flamethrowers* (1989).

Was there a history of Super 8 filmmaking in Germany that you tied into?
No, not at all. In fact, I do not know of any history of artistic Super 8 filmmaking here. There had been singular, isolated attempts of making use of this format in an artistic sense, Werner Schroeter's 8mm of the late sixties for instance, but it was only fifteen years after Kodak had introduced Super 8 as a consumer medium that a significant number of makers thoroughly explored and exploited its artistic potential and freed it from the cramped confines of the living room.

What German screening venues and festivals were responsive to screening Super 8 work?
As in the origins of the sixties filmmakers' collectives, it was mostly the producers themselves who were concerned with the exhibition of their work and who started to run cinemas and to found festivals on their own. Due to its fringe status, Super 8 was not limited to the big cities. This is why in many smaller German cities film workshops, collectives, microcinemas, and festivals emerged. Artists groups such as Schmelzdahin, Alte Kinder, Anarchistische Gummizelle, Kober & Döbele, and the like put places like Bonn,

Bielefeld, Düsseldorf, and Stuttgart on the Super 8 map. This resulted in film critic Dietrich Kuhlbrodt stating that "The centres (of adults) are being abolished." Due to Super 8's youthful mobility, it could pop up all over the place: at run-down squats as well as prestigious art schools. Cinemas like the Munich *Werkstattkino* or the Berlin *Eiszeitkino* and festivals like *experi & nixperi* in Bonn or the *Osnabrück Experimentalfilm Workshop* (nowadays known as the *European Media Art Festival*) were forerunners here in terms of the regular presentation of contemporary Super 8 films. In 1988, the Oberhausen International Short Film Festival devoted programmes to the work of Schmelzdahin and Alte Kinder, which must have been one of the first presentations of Super 8 films at that prestigious event.

Do you remember what equipment you used for your filmmaking?
Well, most of our equipment we inherited or borrowed from our fathers who had either lost interest in making home movies or were about to shift to consumer video instead. I loved the simplicity and Dieter Rams' sophisticated design of my Nizo S 30. I guess the appropriation of the outdated equipment of our fathers was a crucial moment for that new movement we were part of: it was as much an attack of the younger generation against established modes of representation and the patriarchal gaze manifested in them as it was a rejection of the rules of the user manual and the advice of how-to books. This was what our collective name, Alte Kinder (i.e. Old Kids) was meant to express: having been indoctrinated by moving images like no other generation before us, we were both media-savvy while trying to explore our medium in some sort of childlike recklessness. We used our fathers' machines, but we did so in a different, transgressive way. ∎

LBRTR: Interference and Periodic Objects, *1999 Exploratorium After Dark #1. (L–R): Keith Evans, Christian Farrell, Jeff Warrin.*

silt

KEITH EVANS, CHRISTIAN FARRELL, AND Jeff Warrin formed the Super 8 filmmaking group "silt" after meeting in an avant-garde film history class at San Francisco State University in 1986. Given the choice of writing essays for grades or making films, they chose the latter. Impressed with one another's work, the trio quickly joined forces. They were attracted to textured, decaying images similar to those being produced by German collectives Schmelzdahin and Alte Kinder, and quickly began experimenting with ways to affect their films in a similar manner. They created pinhole cameras out of Super 8 cartridges, and buried their films in the ground, soaked them in bodies of water, and even hand processed their work in grapefruits. The group further transformed its films in projection, manipulating projection speed, overlaying images, and creating live special effects. These experiments lead to elaborate projection performances, including a showcase at the Sundance Film

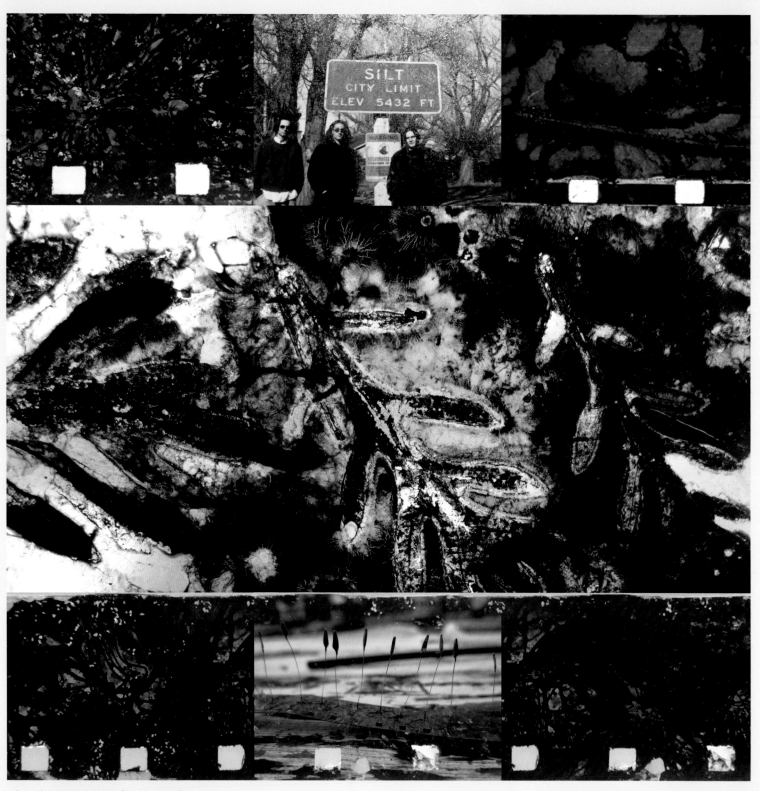

silt with various Microfossilographs.

INTERVIEW

> "Super 8 was essentially so much about chance because we were constantly getting broken cameras from thrift stores, broken projectors from thrift stores. Often what you shot was just an extension of your arm. We were like, 'Point it in that direction!'"

Festival's New Frontier program in 2003. Silt embraced chance and chaos, two factors built into their thrifted Super 8 equipment arsenal. Currently Evans and Warrin live in Bolinas, California. Farrell lives in the Czech Republic.

Can you talk about how you all found each other.
Christian Farrell: We all met in a film class at San Francisco State University.

Keith Evans: I was in conceptual design, which was basically the technological application to the arts program at that time, which was mostly conceptual art.

Jeff Warrin: I was bouncing around in different art mediums: sculpture, photography, taking some film classes. This was an avant-garde cinema class with Bob Bell, a great enthusiast of experimental film.

Was it seeing films in that class that pushed you all into a more avant-garde frame of mind?
KE: I already had a couple of film classes at Northridge in Southern California. That class didn't try to open up the tradition of experimental film. They just gave you cameras and said, "Here, go shoot chickens and we'll teach you how to do it." When I came to SFSU, they had a survey class on this tradition, that was exciting to me.

JW: A few years previous, I remember seeing *Stranger Than Paradise*, Jim Jarmusch's film, and walking out of that movie saying, "I want to be a filmmaker." That was about as experimental of a film as I could imagine. It really was going to that avant-garde class that was mind blowing.

KE: [We] recognized this affinity in the way we were making images and putting things together and we were inspired by things we saw from one another. That just connected us.

CF: I remember Jeff's film, *Journey to a Powwow*, which was shot out of the window of a car and documented the ceremony. We projected it at six frames a second. It was completely hypnotic, and became part of this vocabulary when we started to work together.

KE: The slow motion thing goes back to the home movie scenario. Those old Regular 8 projectors that were variable speed, we would always crank it low and watch those old movies slow. Jeff was shooting with the shutter in a different way and it made these long blurs. When we came together and started looking at our old home movies, those were the sequences we were most excited about—when the colors were moving in painterly ways.

JW: It outlines how much the technology itself determines the aesthetic. That was always the interesting thing with Super 8. The 16mm projectors were pretty standard—18, 24 frames per second, sometimes 16 frames. You were stuck, unless you got into optical printing. Early on, we started started to find these [Super 8] projectors [with various projection speeds].

KE: One Super 8 projector you had was the Moviedeck. That was this important part of like, "Oh, we can do this." We started to see about buying more Moviedecks. They also had a friction feed, so it was different way of driving the film.

JW: We could have film with more crap on it and this machine pulled it through. We later came upon those Eumigs, but Moviedeck was the entry into what we could stuff into a projector and how we could slow it down.

KE: Another aspect of the Moviedeck is you could pull out this little consumer ground glass and mirror object and you could watch it like a little TV in front of you. You [could] see the frame in a different way because you could still them and slow them down. It was a cool way to participate with the Super 8.

CF: Super 8 projectors are like a microscope. You can create a kind of typography on the film emulsion and the projector blows that up in an amazing delicate way. That became this portal we went into. A formative moment was when Keith had been to a performance of Alte Kinder and Schmelzdahin. You had been describing what you had seen, and it really struck me. That day I went home and buried some film in the ground. I didn't know what they had done to get those effects, but it sounded like one could decay film on their own and increase the decay time by putting the film into different environments.

KE: Matthias [Müller] had come with a briefcase filled with these new German Super 8 films. The Schmelzdahin film was otherworldly because it was so abraded. I think it was *Stadt in Flammen*. It was like architectures that flattened into this classic sort of povera, like Sigmar Polke photographs or Anselm Kiefer surfaces, but with a light. This very industrial look but very luminous and just decayed and destroyed. Some of the frames were all burned out. At that time, that's what was most exciting to me—these older films like David Rimmer and some of that beat stuff in the late sixties. The grain of the film was as important as the image.

Is that another thing that you liked about Super 8, what the film stocks were giving you?
KE: The limitations of Super 8 made it more fun. We had those three stocks of black and white and at that point they were phasing out a number of the color stocks. Mostly they were starting to recombine all the Ektachrome into that one G stock. Christian actually had a Fuji camera, which was cool because the Single 8 stock was really different. It had a different base. Something that we found with a lot of the environmental experiments was that the base activated the break-up of the emulsions in totally different ways. The fact that [the Single 8] was polyester rather than gelatin made a big deal.

CF: The way the inks worked in [Kodachrome] was so different than the Ektachrome or the Fuji. The reds would start to float, the blue would come out. Eventually it would all turn blue if you left it in the ground too long. The other ones had layers that would erode, and you'd get a different kind of decaying. Pink, strange colors…

KE: The Fuji had a more pastel spectrum. It would stretch in totally different ways. The environmental stressors would bend the form a little bit or bubble it up. We were excited about this textile quality. In the Fuji films, when it would break down, sometimes you could see a weave rather than a blur in some of it.

You were burying your films and introducing environmental effects. Can you talk about leaving a large part of the filmmaking process to chance?
CF: Collaboration itself created a lot of chance. No one person can completely direct the project. Everybody is bringing in something new that causes a different direction to happen. And then these processes become other aspects of the collaboration. We start collaborating with the place we're working with and inviting chance in that way.

JW: In order to take [a] chance you have to give up your control. The first collaborative project we did…What was the name of it?

CF & KE: *Conflict*.

JW: We created a strategy that we would each shoot the whole roll and the next person did something to that roll, and then passed it around.

CF: Eventually it passed through each of us.

KE: Which meant like reshooting it. So we ended up having more than three rolls of film as the material that we edited.

JW: It was interesting as a rite of passage. First off, this is going to be conflict, so let's title it that. Can we get through this process of letting go?

KE: Super 8 was essentially so much about chance because we were constantly getting broken cameras from thrift stores, broken projectors from thrift stores. Old stock that you'd buy at thrift stores. The eye pieces never totally corresponded to what you shot. Often what you shot was just an extension of your arm. We were like, "Point it in that direction!"

JW: As opposed to having a nice 35mm camera that's capable of different options of nuance, by accumulating fifty different cameras and a hundred projectors you create your own nuance of broken down equipment. Once you start to catalog those imperfections, you've got this palette to draw on. That led to the other part of our process, which was reprojection, like three-dimensional optical printing. We would set up these projectors, and put loops in different ones. We knew different projectors would do different things. Then we'd create this kind of sculptural optical printing setup, where eventually all the images are going to end up on an 18" x 12" piece of vellum paper taped to glass. But how we would bounce projections off of water and put objects in front of the lens and use fire and smoke and all these kind of things came out of getting to know these projectors. There's a communication between the film that you're running through that projector and the qualities of that projector. It becomes like an instrument that has that tone.

CF: Up in the Sierras in the cabin, we'd be down in the basement making *Kuch Nai* after '92, and there'd be multiple loops going up onto the ceiling, reflecting off of mirrors, underwater, and so forth. It would be like jamming, playing these instruments. Things would happen spontaneously in unforeseen ways. That's what eventually lead to wanting to do that live. To create a performance situation in which we would be surprised along with everybody else.

JW: *Kuch Nai* is an example of a film where we created these apparatuses to end up on a single frame in a film that can be played at a festival in an Elmo. But at a certain point, we realized how much more interesting what we were doing behind the screen was to us and maybe to others. That broke us out of that single screen idea.

KE: It was like us being inside of the optical printer. Then we're like, "Let's just do it live."

JW: Which is an interesting catch because dialing in how much you want people to watch you doing it, and how much you want them to actually focus on the two-dimensional output is a fine line.

KE: When we moved into the cinema with all the equipment, suddenly people were watching you. But you had spent months and months making the images, and they weren't watching that.

You were hand processing your films as well?
CF: We've done a lot of hand processing using traditional chemical methods and also unconventional techniques incorporating mold, bacteria, soil…putting film inside grapefruits…processing with coffee and citrus…

JW: People refer to silt sometimes as the alchemists of Super 8. That's flattering, certainly, but at the same time, none of us are very scientifically minded. All our hand processing, all our burial stuff–there was an empiricism of, "Okay let's try this and see what happens." But there were not really formulas and understanding of chemistry. I met Jürgen Reble, from Schmelzdahin, who was well-schooled in the sciences. He was a chemist. I remember having a conversation with him, and he just pulled out a piece of paper and he wrote down a chemical formula. I never knew what to do with that. Who knows if that's our California looseness or what, but it was pretty wide open.

Can you talk about the editing process?
KE: Editing is the key to all of it. We started doing it in the late eighties and the early nineties and there was just copious amounts of film. It cost nothing. We would just shoot and shoot. We would shove all the reels we didn't use unspooled into bags. Then we would experiment with this unspooled mass of film. We were just pulling it out and being like, "Oh, this is really cool, I'm gonna use this! The material itself was just garbage. It was part of the ephemera of this disappearing medium.

CF: When it came down to editing, sometimes we'd be editing up to four, five different things that were going to be going on all these different projectors to create multiple projections. So we would edit something so that there was a particular image on top of this other particular image, and then it would be cutting to another. Some of the editing would be actually the film material itself. And then part of the editing would be live switching from one projector to another. So we might have a couple of Super 8 projectors on top of each other. Those two would turn off, and we'd go to a 16mm projector on its side filling up that same space, and that would be edited. A particular reel that was edited a certain way, would then lead into another Super 8 diptych. Sometimes the editing got pretty complicated, but we had to keep in mind all these different projectors were going to be turned on at different times during the course of the projection.

JW: It's comical to think now how you can sync things up on your iPhone while you're laying in bed. But doing all this stuff, there was so much room for error, and yet at times we would try for that symmetry. That's part of the reason we had to perform these things. You just couldn't press a button and have it sync up. The editing was kind of trying to allot for multiple images, multiple projections, the speed change, finding moments where we can kind of catch up– improvisational tricks like, "Oh, here's a place where we can slow this image down while everyone else catches up."

CF: There were moments where we pass a lens, or leaf, or an object in between the projector and the screen. That became a way to create a space within the editing, in which to have a physical movement and shadow play incorporated into the projection.

Did you guys do a screening at Sundance at some point?
KE: We did it in a giant catering tent there, which was basically the sort of culmination of all of our field effect studies. People were attentive and I think they'd never really seen anything like that. We were lucky enough to be given the opportunity to do this strange spatialized cinematic event.

CF: Projecting through a giant block of ice.

KE: Twenty projectors, thirty projectors. A big giant pool of water with a scroll.

JW: Sundance took a big risk allowing us to create a physical space outside the theater. We were pretty adamant that the screening room was not the place for our films.

Are you still working with Super 8?
CF: I continue to shoot Super 8. [I use video], a phone or something to explore to a certain point, and then come in with either Super 8 or 16 to play with it further. I've had some screenings here in Český Krumlov,

INTERVIEW

almost always in connection with live music and projecting.

KE: I use it. I shoot and still do some work with the environmental aspects of it. Often, I use it within video performance, where I have video cameras reshooting the images or reshooting the projector. For me it's an extension of our work in the sense that I try to do more than my two hands can do. That's part of the narrative that I look at back through our work. It was all somehow about these kind of translations and translating of the processes, and these ideas, and my body into the image.

JW: There was a point in our evolution where we started to incorporate 16mm. Throughout the rest of the time we worked, we would play with it to varying degrees. I think that the relationship to it was always a bit tenuous. I don't feel we ever did with 16mm what we did with Super 8. Some of that goes back to the gestural quality of [Super 8]. When we discovered that Super 8s just fit in your hand in this way, that you can naturally start swinging them, and putting them under things, and above things, and the relationship to your body as something on a hand, as opposed to this Bolex, which is affixed to your chest, was so key in freeing us up. When we combined that gesture with slow motion, I mean basically silt was created. That was the embodiment of it. ∎

Norwood Cheek

NORWOOD CHEEK STARTED making films as a teenager in North Carolina. In the early 1990s in Chapel Hill, the Merge record scene was kicking into high gear and Cheek was there to document it. He had his own band, Sex Police, and began making Super 8 music videos for his group and for other bands, collaborating numerous times with scene torch-bearers Superchunk. The videos found their way to MTV's alternative rock music show, *120 Minutes*, which aired late on Sunday nights. In 1994, Cheek started the Flicker screening series, an ongoing, monthly festival that didn't charge entry fees, thereby creating a lower barrier to entry for artists. Though Cheek started Flicker in Chapel Hill, festivals sprung up in over a dozen cities across the country in the late 1990s. The *Flicker* zine, which Cheek referred to as "Your Guide to the World of Super 8," was filled with info about labs, film stocks, field reports, and filmmaking tips. Cheek moved to Los Angeles in 1997. He continues to make films, focusing on animation and working as an editor. He still shoots music videos and still slips in some Super 8 from time to time.

How did you come to Super 8 filmmaking?

I had been going to the movies since I was a kid and I loved movies. I always loved *Star Wars*. I saw that when I was ten and it had a huge impact on me. But even with that it never clicked in my head that, "Hey, I could do that." I went to a family reunion when I was fifteen and my cousin herded all of the family into the basement and he said I'm going to show you my movie I made with my best friend. He turned out the lights, bed sheet on the wall, turned on a Super 8 projector, and he had a record that was playing the *Rocky* theme. He and his friends did their own Super 8 remake of *Rocky* that was like twelve minutes. It was the greatest thing I had ever seen. They did all the classic, you know, trying to pretend they were boxers. One of the guys drank raw eggs. As I was watching and leaving the room, it was like. "That's what I want to do." It was fun, it was entertaining, it was collaborating. I asked for Super 8 camera for Christmas and got a Super 8 camera later that year. That would have been '81. It was a Bell & Howell. It was low-fi, no frills. It had a zoom lens, and that was it. It was tiny, but it worked. My first film—it's funny that I'm into animation now because that's what I did first. It didn't have a single frame function so I just pressed the trigger and let go really quickly. I animated this Batman figure along with a Snoopy figure having a battle and Snoopy steals the Bat Copter. The lighting is horrible, but it started this thing that I never stopped doing.

All the while I was really into music. When I finished college in 1990, I was

INTERVIEW

still making Super 8 short films, but had a batch of of Super 8 stock in my refrigerator and didn't quite have an idea for a film. My band was playing with all the bands in Chapel Hill and so I talked to my friend Mac of Superchunk and I said, "Let's make a music video." It seemed like the perfect way to combine two of my passions—film and music. So I did a music video for Superchunk shot on Tri-X black and white. One of the tough things for making short films is coming up with ideas and so music videos give it a whole other life. That song is your story and you can just create some fun and interesting visuals around that. The audio aspect of filmmaking, to me, is always one of the most challenging. You can watch the most pristine and beautiful 35mm film, but if the audio is really bad and out of sync, you're not going to keep watching. With music videos you take away that difficult aspect of figuring out the audio.

I started doing music videos for every band in town and came up with this idea to do a documentary through music videos on the Chapel Hill scene. MTV's *120 Minutes* showed some of my music videos, all shot on Super 8, and the word organically spread. I started doing videos for bands all over the country. It opened the door for me to experiment with shooting so many rolls of Super 8. If you were doing a project on your own, you'd never be able to afford it. How would you ever shoot ten rolls of Super 8 for your own little movie? Maybe once a year. With the music video, there was a budget. The budgets were tiny, but at least there was a budget that covered the film stock, the processing, and the transfer. My Chinon Pacific 200 was one of my favorite cameras I used. It had a built in dissolve function.

I had this whole arsenal of cameras that fit whatever situation I might need to cover whether it was time lapse or doing a dissolve. My other favorite camera was a Eumig Nautica that could go underwater. I loved the fixed focus wide angle lens on that camera. I shot this Superchunk video, *Mower*, and I shot almost that entire video on the Nautica. Everyone responds when you bring out a Super 8 camera. The tone of the room changes. It's like when you're at a gathering and an alien walks in. Heads turn and people's demeanor changes. And musicians—these are people I relate to. They fawn over cool old guitars and [when] you bring out a Super 8 camera, it becomes something to break the ice. Nowadays it just stands out [even more]. I just did a music video for The Beths and we shot some Super 8. They really loved and responded to the Super 8. Nowadays [with] photography and filmmaking, a lot of people just equate to using their phone. It's disposable. You don't even look twice at what you've done. But someone with a Super 8 camera or 16mm camera, it's going to change the whole dynamic in a positive way. Whoever's the filmmaker, whoever's the actor, whoever's the person in front of the camera, they're taking what

they're doing more seriously because it takes money. It's not free. There's so many steps to doing it that everyone steps up their game.

Let's talk about the impetus for starting the Flicker screening series.
I started doing a Super 8 class at the Art Center in Carrboro, North Carolina, and I was still making short films. I was getting frustrated with entry fees to festivals because I was spending maybe a $100 on my short films, then these entry fees to festivals [were] $20, $30, $50. You're not even guaranteed to get in. I was doing a music video for this band Five-Eight in Athens, Georgia and met this guy Lance Bangs. That night we went to meet his friend Angie Grass who was running Flicker out of the 40 Watt Club in Athens. She had taken it over from one of the Pylon guys. She had a Super 8 and a 16mm projector and just showing these short films. I loved the idea of it. Back in Chapel Hill teaching my Super 8 class, I decided I wanted to do that same thing. I want to have this festival that is ongoing, that happens every other month, not just once a year. There's no entry fee. The only requirement is it has to originate on film and it has to be under fifteen minutes. I talked to my friends who ran this club called the Local 506 and they were into the idea. I decided to call it Flicker with the idea that one day there would be Flicker festivals all over the world. I kept the same name and started doing this guide book for filmmakers. It was a hit. It was everything I had hoped it would be. I started getting submissions from all over the country. What was great was that people in my Super 8 class had an outlet for their films in their hometown,

INTERVIEW

Melinda Stone

so their friends got to come in and watch their films. Several of my students like James Parish moved to Richmond and he started Flicker there. David Teague moved to New York City. He started Flicker there. Cory Ryan moved to Austin and she started Flicker there. For a span of about five years there were probably twelve to fifteen Flicker festivals going on around the world. In 1997 I moved to LA and continued doing Flicker here and Roger Beebe took over Flicker in Chapel Hill. It became this beautiful, organic thing and then The Attack of 50 Foot Reels became the offspring of Flicker and that started outgrowing and outshining Flicker. I [did] that up until 2010.

What kind of films were you screening?
It was full range of films. There was a lot of stop frame animation inspired by Norman McLaren. Time lapse stuff. One of my favorite local filmmakers, this guy that was more of an engineer, had taken apart a Super 8 cartridge in the dark and masked the gate into four sections. He shot his film four times, reloading the film each time. That film was incredible. One of my favorites was this home movie that some friends made from their honeymoon trip to Las Vegas. The neon signs all looked so great. This is all projected film. It's not transferred to video. What I loved about it was there would be out of focus shots or really shaky camera, but they were narrating the whole film, and they were watching it for the first time themselves. It was so real and candid. The camera would be out of focus for a few seconds and the guy was like, "I must have given you the camera then." Their banter was incredible. I realized then there was this performance aspect of Super 8 filmmaking and film screening. I still go to the movies and I love all types of filmmaking, but I think there is a point where a lot of modern filmmaking, like with digital photography, is that things can get scrubbed clean and crystal clear and perfect that they lose part of their soul. And I think that's what I always have loved and still love about shooting film. There is this dreamlike quality. There's the unpredictable aspect of it. It's an organic thing and I think that gives it heart and personality and makes it unique.

Can you talk about the *Flicker* guide?
I had gathered so much information from all the magazines I collected. I knew every lab and what services they offered. I was fielding so many questions I decided I'm just going to make the *Flicker* guide that has all this information. Angie Grass had a version of that as well. I would update it every year. One of my favorite things was doing the programs for the Flicker screenings. I would put together a program that would have articles and I would enlist some of my favorite filmmakers to write articles about building your own frame counter with a calculator, little cool mods you can do. The *Flicker* guide of course was pre-internet so it's not like I could just check on the Web to see if these companies still offered these services. I would always have to call them, which was a long distance phone call. I loved selling them too. I felt like I was selling a newspaper. That's the only way you could get that information, through these kind of catalogs or zines. The Brodsky and Treadway *Little Film Notebook* was the closest thing that was a resource guide. One of my favorite things was the "Manifest du Flicker." It was a manifesto. I railed against a hundred million dollar movies. That's when movie budgets were going up so high. I thought there should be a cap on movie budgets. You have so many creative filmmakers trying to make it on their own, scraping by, trying to save enough money to enter Sundance, and then you have one movie spending over a hundred million dollars. I thought it was obnoxious. I still think it's obnoxious. If you can't tell your story in a compelling and interesting way for under a hundred million dollars then something is wrong with you. ■

(L-R): Melinda Stone, Kate Haug.

MELINDA STONE HAS always approached film from a unique angle. Collaborating with Igor Vamos of The Yes Men, she created the film and book project, *Center for Land Use Interpretation Photo Spot Project* (1996), which took inspiration from the "Kodak Picture Spot" program that placed signs at picturesque locations across the country, encouraging tourists to snap family photos there. Stone and Vamos turned this campaign on its head by placing their own "Suggested Photo Spot" signs in less traditionally photogenic places like abandoned oil drilling sites. In their infamous Barbie Liberation Organization prank, the duo swapped the voice boxes of Barbie and Ken dolls, and placed them back on store shelves in time for Christmas. In addition to making films, Stone has always focused on creating collective experiences. In 1998, she founded the Super Super 8 Festival, a travelling exhibition that screened at media arts centers, microcinemas, and oddball exhibition spaces and featured live musical accompaniment, Bingo games, raffles, and sing-a-longs. Another Stone project,

Drive-In slides from Melinda Stone's California Tour (2003).

INTERVIEW

Fleur Power (1998), involved planting white flower seeds on the side of a hill, then returning to use them as a projection screen months later when the flowers bloomed.

Stone started the Film Studies Program at the University of San Francisco in 2005. She has since moved on from film into the field of Urban Agriculture.

> "Part of the tour was me reaching out to the venues months before [and saying], "Find us a band that will play with these films."

What drew you to film as an artistic medium?
It was my first year of graduate school at UCSD. I was twenty-seven years old and had been doing journalism for a couple of years and was disenchanted with that style of storytelling. I worked at an NBC television affiliate at the time. There were a lot of constraints put on what I was able to do as a journalist. I took a few years off, and then went to graduate school, interested in studying the sort of "bad media." In the first week I was there, I was on a tour of the film facilities, looking at some films that they showed and seeing all of the film equipment that was available to me as a graduate student. I ended up in a 16mm film production class. What drew me to film was this realization that I can tell any story I wanted to. Looking back on it now I'm thinking, I was twenty-seven years old when I figured that out. What's beautiful now is people figure that out when they're much younger, but I'm thankful that I figured it out.

You mentioned you started with a 16 class, so I'm curious how you got to Super 8? Also, the Barbie Liberation films, which I associate with you, are video. How did you move from medium-to-medium?
I can't say that there was one defining moment where it's like, "Super 8 is really what I want to do!" A lot of it just had to do with being interested in the democratization of media and storytelling. Super 8 represented that with film. It was easy "plug & play." Living in San Diego at the time there were twenty Super 8 cameras at every thrift store, and most of them worked. I started teaching kids at the San Diego Museum and it was much easier to teach Super 8 filmmaking than it was 16.

Do you remember when you made your first Super 8 film?
Doing the Super Super 8 Film Festival I did sing-alongs. I think it must have been a sing-along. I really liked audience participation and it was a way to deal with some of the silent film stocks.

Do you want to talk about founding Super Super 8?
It really was like, "What are people doing out there?" When Igor Vamos and I were traveling around and connecting with all these media arts centers all over the country and connecting with different people, I saw that there were these wonderful havens or oases of media happening all over the place. Having correspondence with people like you or Bill Daniel, people I'd never met, but were passionate about what they were doing, that's what I was drawn to. I found it more with the folks who were doing Super 8. There seemed to be this underground zine like thing going on with Super 8. There weren't any film festivals that were exclusive to or actually had a space for Super 8. At the time, [I was studying] and writing about amateur film. That's folk film culture and I think Super 8 also represents folk film. It all fit together for me. Because I had all these connections with all these media art centers and I wanted to keep traveling around the country, it was super easy to start that film festival. There were things like the *AEIOU* (Alternative Exhibition Information of the Universe) guide.

The thing about that festival was I was open to all genres, but I had a limit. You couldn't have a film that was over ten minutes, and ultimately I never really had any that were over five. I was a huge fan of having a program that was under an hour. I also was looking for films that wanted live musical accompaniment. Usually we had at least three films that needed live musical accompaniment, and part of the tour was me reaching out to the venues months before and sending them a VHS tape of those films [and saying], "Find us a band or a couple bands that will play with these films." That was fun for me and my touring companions that every show is a little different based on the musicians who were there. We would always have packed houses and it was not just that it was Super 8, but also that the musicians were there and they drew on this other population. I remember all of the venues we went to would always say, "Biggest crowd we've ever had" and "Who would have known?"

Can you talk about why you've always gravitated toward alternative forms of exhibition?
Because I was in graduate school and going deep into film history, there are stories I would read about—the early days of cinema. There would be sing-alongs or there would be games of chance. So I didn't feel like I was doing anything new. That's what they say about any kind of art, right? We're just

recycling the past and putting a slight spin on it in a new context. That's what I did. I'm never going to be able to experience this because it's in the past, but I can experience it if I create it. So some of my first Super 8 films were these sing-alongs. What I understood from the past is that "The Star Spangled Banner" would be a sing-along that they would do at the beginning of every film. That was like in the thirties and forties. It was definitely patriotic tendencies around pre-war stuff.

Were they doing this in the big theaters?
You would do it at the Castro. So I thought, what would that be like if we had a different anthem and I was like, "Okay, I'm not going to do that, but what's a song that I like a lot." So we did "Oh My Darling, Clementine." Simple ones that people had a familiarity with. "The Bare Necessities" from *Jungle Book*. Just fun things. "Downtown" by Petula Clark.

What did you film for those? Would you just film the words?
Super 8, either color or black and white. Film the words and sometimes they would have some superimpositions going. It was like live karaoke. Group karaoke, for sure. The music was always on a turntable, and it was always a little off and that was fun. I could slow things down and speed things up. It was playful. It wasn't like, everybody has to get this right because I'm not even getting it right in the back trying to sync things up.

What other live things did you do?
Bingo was what we played. At a thrift store, I bought two boxes of these beautiful old slide window Bingo cards for a dollar. I still have a box of them. I gave the other box to Paolo Davanzo who took over the Super Super 8 film festival. But I have mine just in case I need to play Bingo some time. I did something called "subtle Bingo," which didn't rely on the cage that you would spin. It was all in my head. I would just tell people, "Think about the number or letter combo you want and you will make me say it out loud." People got super into that. It was psychic Bingo. And there were prizes. I had a good rapport with Kodak, so we were always giving away Super 8 cartridges. Sometimes we'd be thrifting along the way and we'd find Super 8 cameras. We could give away a whole filmmaking package. That was fun because you knew you would instigate other people making films.

Can you talk about the different types of venues you went to?
It was everything from a laundromat in Seattle to somebody's family multipurpose room that would get changed into a venue on the weekends to really beautiful art spaces that were nicely funded. In Boulder, we were in somebody's university classroom after hours. Sometimes we were outdoors, and sometimes we were actually part of film festivals. That wasn't unheard of. That wouldn't necessarily be part of a regular tour. Somebody would invite us to do something at that point. We would always stay in people's homes. Ideally we could stay at least another day and get a tour of the town. I can't say that I lost money, but I know I didn't make money. It was definitely a labor of love. But if you were accepted into the film festival, you were paid for your movies. That was something that was important for me.

What type of films were you seeing?
[Super 8] really is the folk film of film, because it was everything. There were home movies, and even though I said ten-minute limit, I would get thirty-minute epic films that were diary films. I was more interested in the experimental films. After a while I got used to certain names I'd get excited about. Like, "Great, I got another Martha Colburn film" or "Danny Plotnick sent something in this year, cool."

You are very much part of the universe of hand-crafted, DIY filmmaking. Can you talk a little bit about that?
One of my favorite stories involves you. When I first moved to San Francisco over twenty years ago, you were working at Film Arts Foundation and asked me if I wanted to teach a class. I said, "I want to teach this class called Self-Reliant Cinema. It totally relies on your hands and hand processing, hand dying, and hand drawing." You're like, "I'll put it on the books, but no one's going to sign up because we're all into digital now." I was like, "This is a different kind of digital—it's your digits." A month and a half later [you said], "That was the class that filled first." It's the craft. It's people wanting to connect again. That first class was really interesting because I had two graduate students from Stanford. I had all these other amazing students, too, but these are people who are paying top dollar to go to film school at Stanford and they're coming here once a week because this is something they're really excited about. It did feel for a while that it really was this movement that was happening. I did feel like I was one of the people at its core, but it's still happening and I know nothing about it. It's a whole different generation and that's exciting.

Of those handcrafted techniques, were there some that you really liked?
My favorite was the crafts side of things, like thinking about a knitting bee or a quilting circle. I loved just sitting around a room—we would have days where it's like here's tons of sharpies and some bleach and some crazy things that stick on. For me it always comes back [to] I actually like people to connect, and this is one way where I feel that connection happens. With most filmmaking there's a ton of connection that happens because it takes a lot of people to make a film, but there's a hierarchy. I'm always interested in trying to figure out what happens if you democratize or just share more of that. I've always been drawn to those filmmakers who wanted to share. ∎

THE 2000S AND BEYOND

Many of the interviews in this book look backward, to a vibrant point on the Super 8 timeline.

Many of the filmmakers interviewed in this book moved away from Super 8 as their filmmaking careers progressed. While this old guard conjures up fond memories of the medium, to them Super 8 remains just that—a memory. But Super 8 is still a going concern. Filmmakers continue to fully embrace the medium. Lisa Marr and Paolo Davanzo began their filmmaking careers at a point when Super 8 was still the cheapest option on the block, and have continued to support and spread the gospel of Super 8 through their endeavors at The Echo Park Film Center. Karissa Hahn grew up shooting digital video, but has incorporated Super 8 into her toolbox.

Super 8 communities and screening opportunities still exist. Festivals such as London's Straight 8 have been challenging filmmakers to push the boundaries of small gauge filmmaking for the past eighteen years.

Super 8 never went away. The stocks have always been available and cameras are still easily obtained on eBay. Kodak introduced negative stocks in 2004, and in quite a shocking development, brought a new Ektachrome to market in the autumn of 2018. As of this writing, the company's new camera awaits release. What follows are several interviews with filmmakers who actively engage Super 8 today. ∎

INTERVIEW

Lisa Marr and Paolo Davanzo

LISA MARR AND PAOLO DAVANZO embraced Super 8 in the late 1990s and never looked back. In 2001, Davanzo opened the Echo Park Film Center, which he describes as "a school of cinema, an artist in residency program, and an itinerant educational model." By happenstance, Marr walked into the center on the day it opened and fell in love with the place. Over time, Davanzo and Marr fell in love with each other. They've combined their shared DIY work ethic, activist spirit, and community-based ethos into a seventeen-year love affair that is downright romantic. Their work continues to actualize the promise of Super 8 as a democratic medium. In addition to offering low cost and free classes for youth in Los Angeles, Davanzo and Marr travel the world, helping communities in places such as Vietnam, India, and the Arctic Circle make stories about themselves.

How did you end up using film as an expressive medium?
Paolo Davanzo: I'm forty-eight years old. I was born in Italy to an Italian father and a Canadian mother, both social justice activists. My love of cinema began when I was eighteen, nineteen. We're talking 1988, 1989, 1990. I was living in Italy during the first Gulf War. I was watching the news and seeing how media could turn things upside down. My friends were talking about smart bombs and no one dying. It was such a "clean" war. In Italy, where the representation in media is more equitable, I was seeing Kuwaitis killed and bombs landing on homes. Bloodshed. I was seeing the power of cinema for propaganda. Growing up in an activist household I was all about politics and social justice and I said, "Wow, why don't I start studying cinema and combine the power of cinema and the power of politics together?"

My father shot Super 8 as a kid. We are like one of the many where, "Everybody gather in the living room, we're going to watch our trip." But he was bohemian. We weren't watching the Grand Canyon and Disneyland, we were watching images of Kuwait and Venezuela and him as a young man in Paris and Rome. My first camera I bought when I was seventeen at a garage sale in Orange County. When we came from Italy we moved to Orange County.

Do you remember what camera you bought?
PD: I'm not a super techy nerd, but I do remember it was a Eumig. I don't remember the make of it, but it looked really nice. We have different preferences now. We're Nizo freaks. We love the Canon 310. But back then it was just big and beautiful. I remember touching it and going, "Wow, this camera is gorgeous."

As you're growing up and your father's shooting Super 8, is that the mode of home movies back then in Italy?
PD: Totally. Even jumping ahead to video, I remember my father worked in Venezuela for a year and came back in the early eighties with a Portapak. It was a VHS on your shoulder [the] size of a microwave. He embraced it. For him it was a storytelling device. He didn't have the same allure and passion of the flickering image that we all

INTERVIEW

share. For him it was functionality. We made a great film with the young people called *Freetime and Sunshine*, talking to the whole thing about home movies, how it really was about class and privilege. A lot of people of color, their stories were not captured and told because Super 8 was not cheap. You had to buy the camera, you had to buy the film. It did democratize storytelling a little bit, but it still was not accessible for everyone. My father was by no means wealthy—grew up on a farm, first to go to college. He was so enamored by storytelling that he made a considerable investment in his first camera to be able to capture his history. He would hang out in Paris, in the seedy parts of town, and the imagery of the rhythm of the people, of the streets. He did have a painterly eye and a poetic eye. It was purely documentary. On a sound level he had a Uher, the sound recorder comparable to the Nagra. That was amazing because he would sit for hours on the left bank in Paris in the sixties and record singers and poets. This is oral history.

Lisa Marr: My journey is in some ways similar and in some ways different. I grew up in Vancouver, Canada. My grandfather shot Super 8 and Regular 8 of his travels. He was in the navy for his whole life. When he retired, he and my grandmother would travel on these package tours and they'd always be filming stuff. We'd watch those. My parents had a Regular 8 camera and shot a lot of home movies of me as a baby. They have films of their wedding and they actually made some weird narrative films. My parents are not artistic at all. They are kind of lower middle class people that moved into the middle class and were proud of that. When they first got married, they made these little movies about them as newlyweds, that were strange narratives. There were weird jokes about her making horrible food and him having to eat burnt toast, and him going off to work, which is so weird because my mom is antisocial and nervous about expressing any kind of creativity. My dad is very social, but they're not people that go to the theater. They're regular average people. They got married in 1962. People of our age associate their childhood memories with the look of Regular 8 and Super 8. There's a certain color and look to those things and tone and patina that resonate with people.

I grew up to be a musician. I played in an all-girl punk rock band in Vancouver. It was called Cub. We did a lot of West Coast touring. High 8 was the medium that we captured that era in. At a certain point I ended up moving to LA. I was in another band called The Beards with Kim Shattuck of The Muffs. We were doing a record and we wanted to make short films to go with all the songs. In Joshua Tree, I found a Bell & Howell camera at a thrift store for $15. It was a super basic camera. It had sort of the wood grain tape on the side. It's like the "Woody" of cameras. I made this Super 8 film for this record and I shot some footage out in Joshua Tree. The record was released early 2002. In late 2001, I was living in Echo Park, and I was walking around my neighborhood, and saw this place that I hadn't noticed before, and went in and there was this man there, and he said, "My name is Paolo, and this is the Echo Park Film Center and we're opening today." I said, "Wow, that's cool." So I checked it out and I loved it. And I just made this film, and the record was coming out. And these people said, "Oh, ask that guy Paolo at the Film Center if we can have our record release and show our film." I asked him, and he said, "Sure." So we had this premiere, and other people had made some Super 8 films to go with that record too. That was my introduction to filmmaking and to Paolo and to the Center. Now it's seventeen years later. I like to say I first fell in love with the Film Center because of Super 8, and then I fell in love with Paolo. It's been this journey ever since where we've been sharing our love for Super 8 and community cinema, both in Echo Park, but also around the world. It's been this incredible, beautiful journey.

We're all relatively the same age and a lot of our contemporaries were moving on from Super 8 or even going straight to 16. I'm curious why you stuck it out with Super 8?

PD: We're still one of the few media arts centers, youth community education centers, that shoots a ton of Super 8. It's on the forefront. We are actively keeping it alive, doing hand-processing workshops. The love affair has never ended. Even though Lisa is my partner for life, Super 8 is like the first kiss. You'll never wash that kiss off your face. On a romantic level, that's why we've kept it. On a practical level, we love the feel, we love the texture. When we started the Film Center in 2001, you could buy Super 8, I would say 8 dollars a roll. You got a camera for $10 at a thrift store. A viewer for another $10 or $20. You could process it for another $10. For about $50 you could be a filmmaker. I couldn't afford a computer in the early 2000s. I couldn't afford a digital camera in the early 2000s. But I could pay $50 and make a three-minute movie. It democratized storytelling. Now, you can get a free phone with a phone plan that makes videos. So the way we tell stories is different, but for us it's never been one or the other. Super 8 is still somewhat affordable; it's still accessible. It's still relevant in our lives.

LM: For me it wasn't a necessity, it was a choice. I was interested in Super 8 because I like the look. I like the handmade quality. For a lot of people that's the appeal. We want to work with our hands. We want to have contact. We want to meet other people that also feel that way. It's a tribe. It's important to me to work that way. To work more poetically. To work on a smaller scale. To be able to control the means of production myself. To learn from others. The community is so generous about sharing their tips and tricks and secrets. It's magic to sit in that very intimate scale and watch a film together. That's what I was drawn to. It wasn't

INTERVIEW

Filming in Hanoi.

like, "Oh, I burn to make films, but this is all I can afford."

With Echo Park you're creating a community, and as you go around the world with your Super 8 workshops you're also seeing other communities. I'm wondering if you can share a story about going to some other part of the world and finding this tribe and forging this connection over Super 8.

PD: Since 2010, we've been doing international engagements called *The Sound We See*. We go into a community. We don't make films about a community, they make films about themselves. We give them the resources to tell these stories. It's in the classic city symphony genre. They're black-and-white films. We're in a community for as little as one or two days, for as long as four months, and we're making a twenty-four-hour film, where each hour of the day becomes one minute on the screen. We've done these seventeen times in seventeen different countries. I'll tell one story. We went to Hanoi, Vietnam. We had a colleague [Thi Nguyen] who grew up in Hanoi. She was a bit younger than us, came to the United States, was doing her dissertation at UC San Diego studying community cinema. Loved The Echo Park Film Center idea. She started teaching and mentoring at the Film Center. Then with her family, moved back to Hanoi and started a place called Hanoi Doc Lab. The history of Vietnam is complex, but in a nutshell, occupied by the Chinese for thousands of years, and then French, and then they call it the American War, not the Vietnam War. The GIs never got north of [The 17th] parallel, so there was no analog culture in the North. In the South, because of the GIs and the war, there were Super 8 cameras left. When we arrived they were documentary filmmakers using video. But they never touched Super 8. They had a profound vocabulary of cinema because they made these powerful documentaries under the mentorship of Thi. When we went there, we brought the resources, we created a lab in our bathroom. We left the Super 8 cameras, we left the projectors. At the premiere they were enamored with the medium. We had shot on Super 8, hand processed, we cut it on Super 8, so there was a print ready to project, but we made a digital version for the screening because it was brighter for 400 people. They insisted on doing a huge loop in the gallery. There was probably a forty-meter loop, because they wanted every single person that came in to touch the film, to see the film, and to understand what they were doing.

LM: That film caused a lot of debate. It was amazing to see how seriously people took art there and what kinds of conversations emerged from the process of watching something together. Some people felt that the images were very nostalgic, and it did not represent contemporary Vietnam, and why would people shoot film, and why was the music so crazy? People were defending it. "No, this is our Vietnam, and this does represent not just our history, but our feeling as young people emerging as artists in this new world." In Vietnam, that was the first time that we had created this lab and left the equipment there, because they had no resources. The project has continued to evolve as we deal with each new community and respond to the community needs.

We went to Old Crow in the Arctic Circle that is a community of about 250 people, the Vuntut Gwitchin Nation in Northern Yukon Territory, eighty miles north of the Arctic Circle. There are only about six of the elders that still speak the language fluently.

INTERVIEW

They're really working to pass that language and the culture onto the younger folks. Part of that culture is also not having direct road access to the village. They've chosen not to build roads. You can get there by very small plane, and it's quite expensive, or you can get there in the summer by the river. Being limited on what we could bring there, Super 8 was the perfect choice for the project. We could bring our tiny projector, we could bring a couple of cameras. We could make a lab there, and work with everybody. It was the fall, it was the caribou hunt, and they were preparing for winter, and chopping wood for winter, and catching salmon for the dog sled dogs, and gathering berries. And yet the community took time out during this super busy time to make this film about the community. Number one, because they want people to see how they live. They want that recognized. Each year the caribou are less and less. They're not coming on the path they normally travel. It's important to document this, and they're using this film as a political tool to say, "This is our way of life, it needs to be recognized, and it needs to be preserved." Also, they use the film to deal with language issues. Normally the soundtrack for these films is an ambient track made by local musicians. But this one included language as an element. It was all hand processed using organic material—coffee and tea and pine needles from the trees and the wild cranberries. That brings Super 8 into a whole new area where it's not just handmade and hand processed, but it's using the ingredients of the place where the film is shot.

Filming at Old Crow.

When you started EPFC, what what was your mission and goal? How has that developed over the years?
PD: It's a combination of the three loves in my life at that time which were education, activism, and filmmaking. The final catalyst was my parents passing. I needed to do something bigger than me. I'd been an activist through my parents. But what was my voice? What was my calling? In many ways it was obvious. LA is so much about filmmaking, but not at all about filmmaking. It's about the power lunches and the Hollywood deals and the limousine. I wanted to create a people's film center that was celebratory and unpretentious, and everyone was treated with the same respect regardless if you had a million dollars or if you had bus fare in your pocket. It began as a cinema school and a retail space, and it's grown since to include artists-in-residency, and an itinerant school bus that travels the country and does free workshops and screenings. But the heart of it has always been storytelling. I am one of the founders, but two other gentlemen, Ken Fountain and Joe Hilsenrad, we all had the vision. Lisa, I consider one of the founders because she walked in on opening day and was there from day one.

LM: It was an instance of my life changing. You walk through a door and your life can change. I think that there's still that magic in the world that we can tap into if we're ready for it. The heart of the Film Center always has been Super 8. We've always shown Super 8. We've always done Super 8 classes. It's transformed a generation of young people to go back to film after being raised as digital babies. When they're making their work using analog film and hand processing, they say, "This changes the way I feel about myself as an artist. This changes the way I feel about my community." It's not just enough to show these films and to celebrate them. We need people to keep making them and that's what the Film Center is about. It's not seeing film as this cute, antiquated, retro thing, but it's a living viable way of making work, and that's what's exciting to us. For people of all backgrounds and experiences, to come together through the process. ■

INTERVIEW

Ed Sayers of Straight 8

STRAIGHT 8 IS A London-based film festival devoted to Super 8 filmmaking—with rules. You shoot one cartridge of film with all the editing done in-camera. You submit your raw footage to the festival along with your soundtrack. Straight 8 processes your film and lines up your sound with the first frame of your image for the festival screening. The first time you see your completed film is along with the rest of the audience. That's pretty awesome. Ed Sayers founded Straight 8 in 1999, and it's still going strong.

How many years has the festival been happening?
Straight 8 began last millennium, just. The first screening was May 28th, 1999.

What was the impetus for starting it?
I'd been planning to attempt to make a short on one roll of Super 8 to be cheap, and to see if some of the inevitable accidents would be appealing. After two years of thinking about it I hadn't got round to it, then people power occurred to me and I asked around twenty friends working in film production if they all wanted a try and so we all did. I finally made my film and accidentally a monster was born.

For a younger generation, what's the appeal of making Super 8 films? It's no longer the most affordable format, nor is it the easiest to use.
In a world of instant everything, the delay factor is very appealing. You don't look at the back of the camera and see what you did; you have to wait. Film is a more emotional medium than digital where even low-end cameras are now so super crisp it almost hurts to look at. And with Super 8, with all its additional grain, you get quite an impressionistic feeling, by which I don't mean necessarily nostalgic, especially with the latest Kodak emulsions available in Super 8. Hearing your precious capture medium whirring through your camera next to your ear, while you hold the trigger down, really sharpens the mind for every single shot.

Can you talk about the limitation of one roll of film, in-camera edits?
Straight 8 is all about the prep and the shoot. You have to plan the hell out of it with only one take of each shot, in story order as you edit in-camera. You have to roll with it as there's literally no going back. It's also all about the premiere. Often a director can barely watch their film by the time its premiere rolls around, because they've seen it so many times in various edit, post-production, and soundtracking stages. With Straight 8, the very first time you see your film is at the premiere with the audience, if it's selected. The best Straight 8 films premiere at the Cannes Film Festival. So you may be in Cannes in a packed 250-seater cinema and whilst you know that our jury has selected your film, you have no idea how it's come out, and whether your soundtrack, which you made blind, works with picture. It's thrilling and not for the faint-hearted. Many people enter again and again, and every year new people decide to give it a go. We've always said, "If you have a Straight 8 in you, we want to see it." Because we never set a creative theme, it's very special when all these stories come back to us from different types of creative people all over the world, in the form of little black boxes full of plastic and chemicals that store light so that it can be played to our audiences in cinemas and online. It's like witchcraft and that's why with twenty years nearly under our belt, we don't plan to stop any time soon. ■

INTERVIEW

Karissa Hahn

KARISSA HAHN GRADUATED in 2014 from CalArts, a school with a strong experimental tradition, featuring a production program that still spotlights analog filmmaking. Hahn works in both 16mm and Super 8; her atmospheric work has screened at prestigious venues such as the Ann Arbor Film Festival, CROSSROADS, Wavelengths (the experimental program of the Toronto International Film Festival), and the Anthology Film Archive. She lives in Los Angeles.

When did you first get interested in filmmaking?
I started making videos in 2003. My best friend and I would transform her basement into these elaborate sets. From then on it just became normal to have a film in the works. In high school there were budget cuts and art was the first to go—and a brand new AstroTurf football field appeared. I wanted to make videos, to make anything really. I was fortunate to have an English teacher who would screen my movies in his class. That became my reason for showing up and for graduating. I often tried to opt out of doing final papers by making videos instead. When I got to CalArts my work changed radically. At CalArts, I was watching a lot of Chantal Akerman, Jacques Tati, and Patrick Bokanowski. I was interested in anything that utilized analog optical effects or was structural. The movies I was making before that were largely inspired by soap operas.

What drew you to working in Super 8? This seems like the path less travelled for someone your age, whose first works would have been digital.
I worked with 16mm at CalArts. I learned the optical printer from Charlotte Pryce and

1_/_ _2 (2016).

experimental film techniques from Betzy Bromberg. I learned about duration from James Benning. When I wanted to make something on the side, I would use Super 8. I suppose a lot of my work is about the frame. Choosing this medium is sort of like picking out the right size canvas.

How does the medium affect the artistic output?
I spend my time in between odd jobs—on the commute to work, on walks, recording stories on this tape cassette recorder. The Super 8 image is the only one that makes sense with these sounds. There is a constant roving sound—the hum, [the] distortion that holds hands with the grain. The clicks of the recorder snap splices into understanding and it just acts as the most proper pairing. The final product often ends up as a digital file, but I swim back and forth between analog and digital during the production.

It seems like you primarily shoot reversal.
I have been shooting Tri-X for the last few years because it is often the cheapest on the

INTERVIEW

all this chaos ensues and I am left with remnants that have been formed by the equipment and the process itself. The Super 8 format allows all of this mayhem.

Do you do all your post digitally, or are you cutting Super 8?
Please step out of the frame is my most recent film, and I did cut it. I have been wanting to cut all my others, but am always on to the next one and never get a chance. You're right, it is a challenge and hard to show around, but for my last piece it made sense. I added in a scene, digitally, where the film melts. I wanted this to be projected live so when you see the image melt you can still hear the projector running. I enjoyed cutting it. Maybe it's meditative to have a task as such. I try to be off the computer as much as possible, so this allows me to sit at a desk and feel like I am being productive.

You seem to work more in Super 8 than 16. Shooting in 16mm gives you more options and sharper images. Why shoot analog, but choose the more limiting medium?
This was just a matter of quick thinking. When I see the price of a roll of Super 8, as opposed to 16mm, I go for the cheaper. I live and work paycheck to paycheck. So, I try to get what I can made in between pay periods. I'm like that kid in the Stanford gratification experiment who takes one marshmallow instead of waiting fifteen minutes for two. I am starting to think about this more and plan to shoot my next film on 16mm. For me Super 8 always feels more freeing. The 16mm has always felt more serious. I currently work at a place doing film scanning and obsess over scanner dirt or the quality of the 35mm films. I don't want that kind of stress when I get off work. Maybe, for me, working in this medium is to be lawless. ∎

menu. It has the fastest turnaround time and I can start sculpting as soon as I arrive home.

Why choose a medium that imposes so many limitations?
The limitations work as a starting point for me. I often work in short-form...manically, obsessively. The Super 8 camera allows me to get a thought shot quickly before changing my mind about it or questioning it to death. I often have one roll of film dictate the duration. This gives me a time frame, a starting point, so that I can then move on to its conceptual aspects. It's like this compulsion. I have to know all details of something before signing onto it. My Super 8 camera and projector are both broken, so this gives it leeway in form. Sometimes the camera shoots only when it wants to, and I see this as doing its own thing. The film often gets jammed up in the gate and shows its frames—its teeth. It coughs, hiccups, and gives a temper tantrum. My tape recorder turns off and on, constructing its own composition. I am meticulous in planning and want to know every angle of how the idea could be perceived, translated, dissected...and then

THE LEGACY OF SUPER 8

This is the new dollar sign in movies...

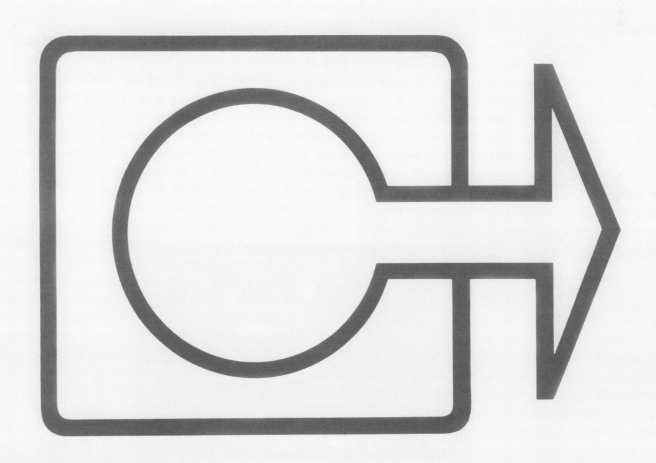

THE LEGACY OF SUPER 8

Over the last fifty years, millions of rolls of Super 8 film have been shot by dads, kids, artists, and Hollywood hotshots.

Due to the challenges outlined in this book, many never bothered to transfer their movies to video. While some films made it onto Beta SP, 1", 3/4", or VHS, many of those tape transfers never managed to cross the digital divide. You would be surprised at the number of classics that can no longer be found. I could give you a list of ten must-see Super 8 films, and I doubt you could find a way to view them without sitting in a room with the filmmaker, a projector, and the camera original film. So, if kids today can't see the Super 8 canon, has it left a legacy?

I believe it has. Super 8's great contribution was its democratizing impulse. It was cheap. It was easy to use. It recorded sound. With a small investment, anyone could start making films. It allowed everyday people, functioning outside the movie industry, to tell stories, document their lives, and chronicle the world around them. Though I'm talking about Super 8, I could just as well be describing today's media landscape and the impulses driving people to make videos on their phones, their tablets, on GoPros and low-cost DSLRs, or with the cameras on their computers.

There are many strains of Super 8 filmmaking whose imprint is apparent in videos currently being produced. Super 8 was a hotbed for diary films. Filmmakers turned their cameras on themselves and used them as confessionals, ranting about their lives, while foisting their opinions and obsessions on the world. Anne Robertson's films may be the defining representatives of this movement. She documented her daily routines, her neuroses, and her obsession with Tom Baker, the actor who played Dr. Who in the 1970s.

Whether or not vloggers like the vlogbrothers (John and Hank Green), WheezyWaiter (Craig Benzine), or Tommy Edison are aware of the works of people like Anne Robertson, it's clear that the urge behind their videos is similar. They vlog about their daily lives and passions, aided by access to low-cost, easy-to-use equipment. Platforms like YouTube allow them to operate and thrive outside the Hollywood system. You can see a line drawn from Super 8 to the vlogosphere. In *How I Feel After 11 Years on YouTube* (2018), Benzine makes the point that his success occurred only after he "let go of perfectionism." This notion of being able to send a film out into the universe and accept that it may not be perfect was the mindset of many Super 8 filmmakers.

The act of direct, unmediated communication lies at the heart of so much material that we see on YouTube and in the podcasting universe. Marc Maron's *WTF* is one of the most popular podcasts going. Each episodes starts with fifteen to twenty minutes of rumblings, grousing, self-loathing, and the laying bare of Mr. Maron's hang-ups. This is the approach Robertson used exhaustively in her films *Five Year Diary* (1981–1997) and *Apologies* (1990).

Anne Robertson's *Five Year Diary (1981–1997)*.

IRA funeral from Les Levine's The Troubles: An Artist's Document of Ulster *(1972).*

Police roadblock at civil rights march outside of Belfast from Les Levine's The Troubles: An Artist's Document of Ulster *(1972).*

Super 8's mobility made it easy to take to the streets. Robertson crashed a Dr. Who convention, hunting for Tom Baker with her Super 8 camera. Scott Stark used his camera to crash Vegas. Filming in casinos is a strictly prohibited activity, yet Stark, with his Super 8 camera in hand, decided to see how far he could get into a casino before being told to stop filming. Stark repeats this act in countless casinos, getting booted out of each one. He compiles it all in his film *11/9/85/Las Vegas/NV* (1985). The film functions as a "super-cut" of these repeated attempts. The term didn't exist back in 1985 when Stark made the film, but this kind of editing is heavily favored by modern videomakers.

Igor Vamos and Melinda Stone's Center for Land Use Interpretation Photo Spot Project *(1996).*

Stark's guerilla spirit is alive and well in filmmakers working today. After Pepsi's debacle of a commercial featuring Kendell Jenner attempting to heal the racial divide with a Pepsi, YouTuber Vito Gesualdi took to the streets to make *Berkeley Protesters Take the Pepsi Challenge* (2017). Gesualdi riffs on the Pepsi ad as he tries to bridge the gap between cops and protesters by offering them sodas. It may not have healed racial conflict in America, but the video racked up over three million views. Gesualdi uses the same hand-held, small footprint tactics that filmmakers like Stark and Robertson used decades earlier. Like Robertson, he films the action, but also acts as an on-screen tour guide, commenting on the world not just with visuals, but through his running commentary. This personal performative approach to filmmaking was a staple of the Super 8 world and is alive and well today.

Gesualdi's pranksterism is representative of much of the work we see on the web. Filmmakers unleash themselves on an unsuspecting public and hijinks ensue. From Greg Benson's shopping list videos wherein he enters big chain stores and asks where they keep unusual items, such as toddler-sized shark cages, to Casey Neistadt's *Bike Lanes* (2011), in which he illustrates the hypocrisy of NYC bike lane laws by smashing his bike into countless obstacles that block his path, to Improv Everywhere's street theatricals, YouTube is filled with videos that bring the artist into close contact with the public.

There were countless Super 8 films mining the same territory. In *Ramp* (1976), James Nares rolls a weighty pendulum ball down an alley in NYC. The ball wreaks havoc in the roadway. Curbsides take a beating. Though the street is empty, we hear kids talking off-screen. A sense of menace pervades as we nervously wait to see if bystanders will be caught in the crossfire. *Center for Land Use Interpretation Photo Spot Project* (1996) finds Melinda Stone and Igor Vamos riffing on a popular Kodak ad campaign. The photo giant placed "Kodak Picture Spot" signs all over the country, overlooking grand vistas, encouraging vacationers to take family photos of these scenic panoramas. Stone and Vamos's version forces tourists to explore the landscape at other points of public interest by placing their "Suggested Photo Spot" signs in locations such as an abandoned oil drilling site in Utah, a sludge depository in Texas, and Kodak's own waste water treatment facility in Rochester.

While pranksterism in both Super 8 and modern video often seeks a laugh, much of the laughter comes from punching up and taking authority to task. This activist impulse can be found in many DIY films and videos. Lenny Lipton noted that Super 8 "was especially attractive to people in third world countries where the low price of the medium was important and also there was no 16mm infrastructure." Julio Neri's *Erase Una Vezen Venezuela* (1977), which takes a personal and critical look at the history of Venezuela, was just one of a handful of films being produced about Third World politics in the 1970s. The act of citizens documenting from the frontlines was the future that Super 8 promised back in the early 1970s. Irish filmmaker Les Levine, who made the documentary *The Troubles: An Artist's Document of Ulster* (1972), summed it up succinctly: "There's a lot of news that we all should be making. Everybody should be making his own news program. Take your Super 8 camera, go out, and make a news program." One of the facets that Levine loved about Super 8 for reporting from the trenches was its mobility. "With Super 8, we were able to run around and get especially difficult shots. I would just tell the camerawoman [Catherine Kanai] to run over to a certain place to see what was happening. We would point the camera, and zap, we got it." Super 8 may not have fully delivered on this promise to create citizen filmmakers, but today's mobile technologies are truly unlocking the goals that Super 8 sought to achieve. Affordable and portable video has led to breathtaking documentation of modern historical movements and political actions, with the Arab Spring, a series of anti-government protests that spread across the Middle East in 2010, marking the coming-of-age moment for this form of video engagement.

"The call to be fearless, open, trusting, and terribly alert is met head-on in small-gauge filmmaking by both artists and viewer." Jytte Jensen said this in 1996, but those words seem even more relevant today. The fearlessness of the citizen filmmaker is absolutely necessary in today's media landscape. Back in 1996, no one could have foreseen how easily marginalized filmmakers would be able to reach a global audience and rapidly disseminate their work. The deep connections that today's YouTubers have with their audiences and the speed at which a video can go viral are wonderful gains that Super 8 filmmakers were striving for twenty years prior.

In part, Super 8 filmmakers tried to forge connections with a general audience by making countless film portraits. They turned their cameras on their neighbors, on their friends, and on people they barely knew. They collected stories from outsiders and individuals. Gary Adlestein and Saul Levine documented artists, filmmakers, and poets who never achieved widespread art stardom. Levine's *Raps and Chants* series (1981–1982), pointed the camera at his creative friends, providing a glimpse into the lives of artists on the margin.

Levine also spent time documenting regular folks. His *Time to Go to Work* (1978) is a portrait of a rail conductor on the Conrail line running between New York and New Haven. This work is updated in contemporary initiatives such as Story Corps, which captures "unscripted conversations, revealing the wisdom, courage, and poetry in the words of people you might not notice walking down the street." This is also an apt description of much of the portraiture work that was done in Super 8.

Super 8 filmmakers documented the conditions in which they lived and worked. In 2005, a DVD box set, *Berlin Super 80,* was released, featuring the work of West German Super 8 filmmakers from 1978–1984. While the set contains many interesting films, what remains seared in my memory is the vibe, the feel, and the sense of place these movies captured. Primarily narratives, music videos, and experimental offerings, the films exquisitely express the milieu in which these artists operated. They were shot in their apartments, starred their friends, featured the music they were listening to, and were art directed with the clothing, furniture, and knickknacks that populated their living rooms. After watching

Gary Beydler's Mirror *(1974).*

this set, you inherently understand the real life of an artist in 1980s Berlin.

Young filmmakers today are also capturing their natural habitats for future media archaeologists to examine. Many YouTubers shoot their videos using the devices affixed to their personal computers. They talk directly to the camera, and in doing so, offer up a view of their lives. We listen to them while absorbing the details of their rooms, the art decorating their walls, the items littering their desks, their clothes, and their hairstyles. These videos will serve as a time capsule of the 2010s; an aesthetic of place and style develops in these bedroom recordings, just like the one archived in *Berlin Super 80*.

Super 8 filmmakers also documented places beyond the four walls of their apartments. They made films about their cities and their neighborhoods, steering away from the touristy spots, and often trafficking in the high-danger zones outside city centers. The gritty, desiccated world of 1970s New York is seen through the lens of Vivienne Dick. In her film *Staten Island* (1978), the borough becomes a science fiction dystopia. Bill Daniel's *The History of Texas City* (1988) turns the camera's gaze on the industrial zones forged by the oil and chemical industries that would be assiduously avoided by the city's tourist bureau.

Filmmakers today are also capturing their neighborhoods, cities, and surroundings in spectacular fashion. Beautiful time lapse videos fill the internet. We see sunrises and moonrises. We watch the stars sweep across the night sky in countless locations around the globe. Many of these videos serve as pure documents, capturing a moment in time like Marsel van Oosten's *Namibian Nights* (2012), a stunning bit of time lapse night photography showcasing the African countryside. Others capture the poetry of nature. Jeff Frost's *Flawed Symmetry of Prediction* (2012) melds time lapse, music, and art into a gorgeous rumination on the American West. These films pair well with Gary Beydler's *Mirror* (1974), which, via time lapse and a giant mirror held by the filmmaker, simultaneously captures an entire day's worth of light traversing both the eastward and westward skies, with the westward view featuring light reflecting off the Pacific Ocean.

Today, there tends to be an emphasis on the beauty of nature, a stark contrast to much of the Super 8 work of the 1970s and 1980s. The brutality of urban spaces was a focus of Super 8 city dwellers, who dealt with the effects of cities on the verge of bankruptcy. That said, today's films aren't all gorgeous. *Poglos Reverberation* (2015) by Radek Sirko showcases bleak Polish landscapes, which serve as backdrops for noise bands interacting with these locales in a deafening way. The austere dreariness on display, in many respects, taps into the desperation evident in films like *Staten Island*.

In the 1970s and 1980s, artistic enclaves were centered in big cities like New York, Berlin, and San Francisco. Films that commented on culture and politics tended to be made in urban centers and seen at festivals and art spaces in those very same environments. Super 8 envisioned a worldwide phenomenon that moved beyond the major metropolises. However, the lack of distribution for small gauge films in the 1970s and 1980s meant that much of the content that wasn't produced in the big cities had trouble finding an audience. Today, thanks to digital distribution platforms, we can readily see films from all over the world. The dream of Super 8 once again comes to fruition.

In one significant way, Super 8 did offer a view beyond metropolitan life. Let's not forget that from the outset, home movies were the driving force of the Super 8 machine. Via the home movie, we got a look into small town America. Countless hours of footage were shot every month by moms, dads, aunts, and uncles. Looking back at these movies provides a glimpse into the world of the mid-1960s. We saw inside the houses of suburban America. We saw moms cooking in the kitchen, the family eating dinner in the dining room, and kids lounging in the TV room, goofing off on the front lawn, on the porch, in the backyard, and bicycling up and down their streets. We saw bouffant hairdos, peddle-pushers, and homemade dresses, pants, and sweaters. We saw wallpaper, shag carpet, wood paneling, and linoleum. We saw graduations, baby's first steps,

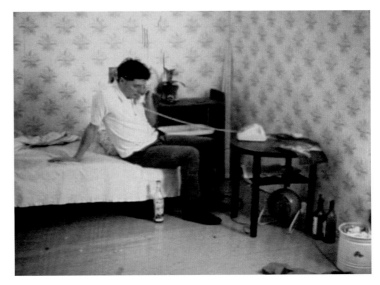

Georg Marioth's Morgengesänge *(1984).*

Bill Daniel's The History of Texas City (1988).

ski trips, sledding, people playing with pets, kids spraying each other with water hoses, and let's face it, a lot of Christmas trees and Christmas present unwrapping. While it would be a stretch to say that watching a family unwrap Christmas gifts in Des Moines in 1967 is the precursor to unboxing videos that clog the arteries of the internet today, it's pretty clear that YouTube is the modern day clearinghouse for all of these heartfelt milestones and home-movie moments. In 1965, these were shot on Super 8. In 2019, they are recorded using cell phones.

Here, a very clear line can be drawn from Kodak's Super 8 vision of a world where anyone can pick up a camera and document life to that vision fulfilled by the technologies of today.

Super 8 also promoted the ease of sharing home movies with each other. Classic Super 8 advertisements from the 1960s showcased the family gathered around the home projector watching movies. The internet takes this vision and runs with it. No longer do you need to be in your family's living room to share these touching moments. As American families spread out across the country, connection to one's family becomes more tenuous. If you live in Phoenix and your mom lives in Chicago, you can still share your child's first steps via online video. We could recreate those classic Super 8 projector ads, but instead of a family gathered around one screen, we would see multiple screens, in multiple locations around the globe, with everyone viewing the same Kodak moment.

The offshoots of the family home movie were the genre films spearheaded by mischievous teens. They made horror movies, comedies, heist movies, riffing on nearly every type of popular movie form. In *American Movie* (1999), a documentary about filmmaker Mark Borchardt attempting to make the great American horror movie, we see his teenage Super 8 horror films. Boasting titles such as *The More the Scarier* (1980) and *The More the Scarier III* (1986), Borchardts's teen films star his friends and take place in neighborhood cemeteries and junkyards. In Redding, California, there was a group of teens pumping out homemade sci-fis and heist movies under the moniker Necro Films. In Northern California, far away from the industry, this group acted like a mini-studio, pooling their resources, equipment, and talent. They literally ran amok in their town and kicked out films filled with energy and passion.

You see the same spirit today on the internet. Often, early missives from the ninth- and tenth-grade set are home movies in the Borchardt vein. The kids pick a genre they like and they start making movies. They grab their phones or the family DSLR, gather their buddies, and try to make the next Edgar Wright or Tarantino opus.

Once again, we see a clear through-line drawn between the early days of Super 8 and the media landscape that we live in today, the barrier of entry lowered so that anyone in the family can make a movie. The democratic beauty of Super 8 was that it put the ability to make films into the hands of the people. And when you put thousands of cameras into the hands of people unbound from the rules, anything can happen.

While we can clearly see how the videos of today resonate with Super 8 films produced twenty and thirty years ago, the reality is that when we think of film history, Super 8 is relegated to a footnote at best. Why? How did Super 8 become a vanished medium?

Though it was cheaper than 16mm, it was still expensive. Though it was easier to use than 16mm and Regular 8, it was still complex. Though it produced a nice image when shot perfectly, it was often shot imperfectly by amateurs and first-time filmmakers, with the resulting images leaving a bit to be desired. For all the talk of democratization, it cost too much and was too hard to use.

Super 8 also suffered from bad timing. It got plowed under by video in the early 1980s. By the time the internet rolled around, even though filmmakers had the ability to archive and showcase their films online, in many cases, this never happened.

Super 8 may have never achieved its dream, but when I think of its legacy, I can't help but reflect on the current media landscape. Today's low-cost, easy-to-use digital video equipment, combined with the power of internet distribution, has transformed who gets to make movies and how we watch those films. This is the vision Super 8 promised over fifty years ago. Super 8's dream is alive and well. And let's not forget, there are many young filmmakers who embrace the medium and continue to write its history. While this book has talked a lot about Super 8's much discussed and debated death, I like to think of it as one volume in a living history—with many chapters still to be written. ∎

Julio Neri shooting on location in the Andes (1978).

NOTES

CHAPTER 1

14 — is almost 50 percent larger than the 8mm image: E. A. Edwards and J. S. Chandler, "Format Factors Affecting 8mm Sound-Print Quality," *Journal of the Society of Motion Picture and Television Engineers* 73:7, (July 1964), 537.

14 — an image with greater definition: Kodak, *Kodak Press Release*, March 1, 1965, 2.

14 — "were surprised at how well the new format caught on": Alan D. Kattelle, *Home Movies: A History of the American Industry, 1897-1979* (Nashua: Transition Publishing, 2000), 212.

14 — Kodak discontinued all but one Standard 8mm model: Kattelle, 212.

16 — "educational, industrial and commercial purposes": Edwards and Chandler, 537.

16 — "possible to make four Super 8 prints": Kodak, "Super 8 Film–Important Medium of Communication in Business, Industry, Education," *Kodak Press Release*, April 4, 1966, 2.

16 — "recognized that it would be of significant benefit": Roland Zavada, "The Standardization of Super 8," *Journal of the University Film Association* 22:2 (1970), 39.

16 — As early as 1966: Kattelle, 212.

16 — nearly thirty companies: Kattelle, 202.

16 — more than eighty companies: Super 8 camera companies listed in Jürgen Lossau, *Filmkameras Movie Cameras* (Hamburg: atoll medien, 2000).

18 — Young filmmakers involved in the German film organization Gegenlicht: Reinhard Wolf (editor of shortfilm.de), email message to author, October 28, 2018.

20 — over eighty American universities: Antonio Sanabria, "Guide to Super-8 Schools," *Super-8 Filmaker*, May/June 1975, 19.

20 — "Bell & Howell closed its Consumer Division": Kattelle, 239.

20 — Kodak stopped manufacturing: Kattelle, 239.

CHAPTER 3

32 — "equipped with a 6 to 66mm Schneider Optivaron f/1.8 zoom lens": Lenny Lipton, *The Super 8 Book* (San Francisco: Straight Arrow Books, 1975), 71.

32 — "It's a fine French machine": Lipton, *The Super 8 Book*, 74.

32 — "nearly went bankrupt": Lutz Auhage and Jürgen Lossau, *Nizo* (Hamburg: atoll medien, 2012), 13.

32 — "taken over by Braun": Auhage and Lossau, 71.

32 — By 1967, there were already over seventy models: Alan D. Kattelle, *Home Movies: A History of the American Industry, 1897-1979* (Nashua: Transition Publishing, 2000), from chart on 202.

34 — there were over 1,600 models available: Super 8 camera models listed in Jürgen Lossau, *Filmkameras Movie Cameras* (Hamburg: atoll medien, 2000).

34 — After all, more than 10,000,00 Super 8 cameras: Kattelle, from graph on 107.

44 — The Pathé also came in: Lipton, *The Super 8 Book*, 76.

44 — The Canon was a steal: Lipton, *The Super 8 Book*, 76.

44 — the release of their Filmosound 8: Kattelle, from chart on 236.

44 — the Bell & Howell system was not reliable: Kattelle, 237.

44 — Section beginning: "That's a quick glance at the whole process": Lipton, *The Super 8 Book*, 133 and Stephen R. E. Aubery, "Hear's the Word on Double-System Sound," *Super-8 Filmaker*, May/June 1976, 19.

CHAPTER 4

56 — According to Roland Zavada: Roland Zavada (Chairman of the Committee on 16mm and 8mm Technology for SMPTE), interview with author, August 14, 2018.

56 — 1988 represents peak film stock in terms of "the number of different films that were made" across all Kodak film: Robert L. Shanebrook, *Making Kodak Film*, expanded 2nd ed. (Rochester: Robert L. Shanebrook, 2016), 375.

56 — "There's always a question of what is the color": Zavada, interview with author, August 14, 2018.

58 — "ten labs that processed Kodachrome," with another "ten non-Kodak owned labs": Shanebrook, 411.

58 — spearheaded by research chemist: Alan D. Kattelle, *Home Movies: A History of the American Industry, 1897-1979* (Nashua: Transition Publishing, 2000), 216.

58 — Gorman played around with the stock's chemistry: Lenny Lipton, *The Super 8 Book* (San Francisco: Straight Arrow Books, 1975), 18.

CHAPTER 5

As noted in Chapter 3 ascertaining accurate release dates for certain pieces of equipment is challenging. This chapter contains such challenges and I have noted where potential discrepancies may arise.

70 — Section beginning: If the scratch occurred on the film's base: Dino Everett (Archivist, USC SCA Hugh M. Hefner Moving Image Archive), word document to author, August 20, 2018.

70 — Section beginning: Nowadays, when restoring a film: Everett, word document to author, August 20, 2018.

72 — In traditional movie theaters: Michael Anders (projectionist Castro Theater, San Francisco), interview with author, August 20, 2018.

75 — The Elmo ST 1200D, which was available for general release in 1976:

Lenny Lipton discusses this projector in *The Super 8 Book* which was published in 1975.

Super-8 Filmaker makes note of its announcement in May/June 1976. *Popular Photography* makes note of its announcement in July 1976.

Lenny Lipton, *The Super 8 Book* (San Francisco: Straight Arrow Books, 1975), 269.

"New Products from PMA," *Super-8 Filmaker*, May/June 1976, 45.

Leendert Drukker, "What's New from the Chicago Photo Show: Movies," *Popular Photography*, July 1976, 128.

75 — The Elmo GS 1200, which hit the streets in 1977:

This projector is previewed by *Super-8 Filmaker* in November/December 1976, noting that it is slated for a release in Spring 1977.

Gunther Hoos, "Photokina '76," *Super-8 Filmaker*, November/December 1976, 41.

75	came with a standard: Hoos, 41.
75	a whopping $3,700 price tag: "Photokina '80: Space Age Super-8," *Super-8 Filmaker*, December 1980, 29.
75	set you back $1,500: "The Great Projector Roundup," *Super-8 Filmaker*, June 1980, 26.
76	"Want to get high on Sophia Loren": Carole Kahn, "Fly Me, I'm Super -8!" *Super-8 Filmaker*, March/April 1974, 27.
77	"By 1974, fifteen international airlines": Kahn, 27.
77	The Trans Com projector: Kahn, 29.

CHAPTER 6

82	The light from the bulb: Robert Price. "Editor/Viewers: The Inside Story," *Super-8 Filmaker*, January/February 1978, 41-42.

CHAPTER 7

90	"corduroy, vinyl, Velcro srtrips": Robert P. Brodsky and Antoinette Treadway, *Super 8 in the Video Age: Using Amateur Movie Film Today*, 3rd ed., 6th printing (Rowley: Brodsky & Treadway, 1991), 46.
90	making a seal skin barney: Beverly Ensom. "So You Think You've Got Problems; Look at Filmmakers of the Frozen North," *Super-8 Filmaker*, July/August 1977, 20.
90	it was used to ensure: Ensom, 20.
92	"find the microphone in this picture": Kodak Advertisement, Kodak Ektasound 240 ad, 1977.

CHAPTER 8

98	"Kodak actually had to": Roland Zavada (Chairman of the Committee on 16mm and 8mm Technology for SMPTE), interview with author, August 14, 2018.
100	"I remember counting": Dave Markey (filmmaker), interview with author, June 11, 2018.
101	All quotations in Polavision section attributed to: Alan D. Kattelle, *Home Movies: A History of the American Industry, 1897-1979* (Nashua: Transition Publishing, 2000), 188.
101	"The size of a large Xerox machine": Chuck Cyberski, "Professional Super-8 Comes of Age," *Super-8 Filmaker*, November/December 1975, 44.
101	"about ten Super 8 cartridges per day": Super 8 Sound Advertisement, *Super-8 Filmaker*, September/October 1974, 33.

CHAPTER 9

108	Section beginning: To achieve optimal colors: Keith Anderson (co-owner of Yale Film and Video), interview with author, September 7, 2018.
108	Section beginning: "Keith Anderson, co-owner of Yale Film and Video, told me": Anderson, September 7, 2018.
110	"One would think that": Ross Lipman, "Technical Aesthetics in the Preservation of Film Art," in *Big As Life: An American History of 8mm Film*, ed. Albert Kilchesty (San Francisco: San Francisco Cinematheque, 1998), 89-90.
111	"[They] used a 16mm print": Anthony Slide, "The Collector," *Super-8 Filmaker*, January/February 1975, 55.
111	"direct descendants of original negatives": Eric Beheim and William Wind, "Sneak Previews: Secrets of a Collector," *Super-8 Filmaker*, May 1979, 55.

CHAPTER 10

	Much information in this chapter gleaned from conversations with Dino Everett, Archivist, USC SCA Hugh M. Hefner Moving Image Archive and Buck Bito and Jennifer Miko of Movette Film Transfer located in San Francisco.
114	The projector was outfitted with a five-bladed shutter: Dino Everett (Archivist, USC SCA Hugh M. Hefner Moving Image Archive), word document to author, January 2019.
117	"What Derek was seeing": James Mackay (producer), interview with author, June 21, 2018.
119	Information in Supermatic section from: Chuck Cyberski, "Professional Super-8 Comes of Age," *Super-8 Filmaker*, November/December 1975, 44 and Super8 Sound Advertisement, *Super-8 Filmaker*, September/October 1974, 33.

CHAPTER 11

126	Excerpt from: Marc Huestis, *Impresario of Castro Street: An Intimate Showbiz Memoir*, (San Francisco: Outsider Productions), 2019, 58-60.

INTERVIEWS 1960S/1970S

136	"Lipton: Tell me, what have you got against 18 fps?": Film quote from *Children of the Golden West*, directed by Lenny Lipton (1975).
137	8mm was perceived as: Gerry Fialka, interview with author, May 23, 2018.

ROLAND ZAVADA

138	By reading the initial 1964 SMPTE article: E. A. Edwards and J. S. Chandler, "Format Factors Affecting 8mm Sound-Print Quality," *Journal of the Society of Motion Picture and Television Engineers* 73:7, (July 1964).
139	I read in the Home Movies book: Alan D. Kattelle, *Home Movies: A History of the American Industry, 1897-1979* (Nashua: Transition Publishing, 2000), 211.
139	In your 1970 article: Roland Zavada, "The Standardization of Super 8," *Journal of the University Film Association* 22:2 (1970), 39.
140	Another thing about standardization: Zavada, "The Standardization of Super 8," 41.

LENNY LIPTON

142	"The communication is now open to people…" Film quote from *Children of the Golden West*, directed by Lenny Lipton (1975).

PAUL SHEPTOW

153	"The prenatal classes": Donald Zimmerman, "It's a Super-8 Baby," *Super-8 Filmaker*, Fall 1973, 30.
153	*Super-8 Filmaker* published 60 issues: Richard Jantz, Editorial, *Super-8 Filmaker*, July/August 1981, 5.
153	*Moving Image* lasted just one year: Alan D. Kattelle, *Home Movies: A History of the American Industry, 1897-1979* (Nashua: Transition Publishing, 2000), 260.

INTERVIEWS 1980S/1990S

174	"Rock 'n' roll bands said": Luc Sante, "Mystery Man," in *Jim Jarmusch Interviews*, ed. Ludvig Hertzberg (Jackson: University Press of Mississippi, 2001), 90.

MATTHIAS MÜLLER

194	In the essay you wrote for the book *Derek Jarman Super 8*: Matthias Müller, "Magic & Loss: Selective Blindness & Analogue Reality," in *Derek Jarman Super 8*, James

Mackay et al., (London: Thames & Hudson, 2014), 144.

THE LEGACY OF SUPER 8

218   "let go of perfectionism": Craig Benzine (WheezyWaiter), "How I Feel After 11 Years on YouTube," YouTube video, 5:29, June 5, 2018, https://www.youtube.com/watch?v=DawA508dLXw.

220   toddler-sized shark cages: Greg Benson and Ryan Smith, "Black Friday Shopping Prank!", YouTube video, 3:59, November 28, 2011, https://www.youtube.com/watch?v=CYbVpAwGGGs.

220   "was especially attractive to people in third world countries": Kathy Geritz, "I Came into an 8mm World," in *Big As Life: An American History of 8mm Film*, ed. Albert Kilchesty (San Francisco: San Francisco Cinematheque, 1998), 48.

221   "There's a lot of news that we all should be making": Joyce Newman, "A New Medium for Avant-Garde Artists," *Super-8 Filmaker*, Fall 1973, 21.

221   "With Super 8, we were able": Newman, 19.

221   "The call to be fearless": Jytte Jensen, "8mm: American Images & the Art of the Everyday," in *Big As Life: An American History of 8mm Film*, ed. Albert Kilchesty (San Francisco: San Francisco Cinematheque, 1998), 17.

221   "unscripted conversations": "Stories," StoryCorps, accessed Fall 2017, www.storycorps.org/stories/.

Kodak, "Chronology: KODAK Super 8 Movie Cameras: 1965-1981," *Kodak Internal Document*, no date, 1-6.

Kodak, "Projector Name and Model," *Kodak Internal Document*, no date, 1-6, 11-19.

Jürgen Lossau, *Filmkameras Movie Cameras* (Hamburg: atoll medien, 2000).

Jürgen Lossau, *Filmprojektoren Movie Projectors* (Hamburg: atoll medien, 2005).

http://super8wiki.com/index.php/Category:Manufacturers

http://super8wiki.com/index.php/Main_Page

https://www.filmkorn.org/super8data/

http://www.bolexcollector.com/timeline.html

NOTES ON EQUIPMENT PRICING AND YEARS OF MANUFACTURE

This information was gathered from the following sources. Please note that these sources often provided contradictory information. I attempted to list information that I could corroborate from multiple sources. Information on cameras is the most complete. For projectors, I often listed only the year of original manufacture. For editing equipment, I could not always find the year of original manufacture, so I listed a date when I knew it was available for purchase.

SOURCES FOR EQUIPMENT PRICING

Equipment & tradeshow roundups from *Super-8 Filmaker*, 1972-1981.

Equipment & tradeshow roundups and camera store advertisements from *Popular Photography*, 1965-1985.

Alan D. Kattelle, *Home Movies: A History of the American Industry, 1897-1979* (Nashua: Transition Publishing, 2000).

PHOTO CREDITS

4   Photo by Danny Plotnick
7   Colored tabs desgined by Ronnie Garver

TABLE OF CONTENTS

8   Minolta Autopak-8 K11, 1966-1968, $179.50.
Chinon 877 Macro, 1976-1979, $349.50.
Bolex 18-5L, 1967-1973, $160.
Korvette Convertible 8ZR.
Photos by Danny Plotnick.

FOREWORD

10   Photo by Danny Plotnick. Design concept by Ronnie Garver. Design by Joanne Chu.
11   Agfa Movexoom 4000 user manual. Courtesy of Agfa-Gevaert NV. Image Source: Super8wiki, http://super8wiki.com/index.php/User_Manual_for_Agfa_Movexoom_4000_Synchrosound_(German)

CHAPTER 1

13   Elmo Super 106, 1974, $151. Photo by Danny Plotnick. Design by Jeremy Troy.
15   Design concept by Ronnie Garver. Design by Victor Krummenacher.
Kodak Brownie 8mm, 1951-1956, $47.50.
Kodak XL 33, 1971-1975, $119.50.
16mm Bolex camera.
Photos by Danny Plotnick.
16   Color film strips. From 1965 brochure, "This Is the New Dollar Sign in Movies…" Courtesy of Eastman Kodak Company. Line art. Used with permission from Eastman Kodak Company.
16   Film strip line art. Used with permission from Eastman Kodak Company.
17   Blueprint of "8mm x 100' Return Reel for Super 8 Film," November 7, 1967. From Kodak Historical Collection #003, D.319, Rare Books, Special Collections, and Preservation, River Campus Libraries, University of Rochester. Used with permission from Eastman Kodak Company.

Invention Report. "Reel with Film Attaching Pin and Slot," November 7, 1967. From Kodak Historical Collection #003, D.319, Rare Books, Special Collections, and Preservation, River Campus Libraries, University of Rochester. Used with permission from Eastman Kodak Company.

Kodak News Release, "New Super 8 Kodachrome II Film Offers Screen Images That Are Brighter, Sharper and More Stable," April 26, 1965. From Kodak Historical Collection #003, D.319, Rare Books, Special Collections, and Preservation, River Campus Libraries, University of Rochester. Used with permission from Eastman Kodak Company.
Collage by Danny Plotnick.
18   Kodak ad, "Load Instantly, 1966." Courtesy of Eastman Kodak Company.
19   Canon 312 XL-S. Photo by Danny Plotnick.
Eumig ad, Mark S-810, 1975.
Agfa Movexoom 4000 user manual. Courtesy of Agfa-Gevaert NV. Image Source: Super8wiki, http://super8wiki.com/index.php/User_Manual_for_Agfa_Movexoom_4000_Synchrosound_(German)
20   Kodak ad, "Already Have a Camera," 1966. Courtesy of Eastman Kodak Company.
21   Courtesy of the George Eastman Museum. Used with permission from Eastman Kodak Company.

CHAPTER 2

23   Photo by Christian Bruno.
26   Photo by Danny Plotnick.
27   Illustration by Jeremy Wheeler.
29   Still provided by Dino Everett, Archivist, USC SCA Hugh M. Hefner Moving Image Archive.

CHAPTER 3

Photos from 31-47 by Danny Plotnick.
31   Bell & Howell Autoload 374F, 1970, $79.95.
48   Courtesy of Pro8mm.
49   Super8 Sound Ad based on layout of Super8 Sound ad from *Super-8 Filmaker*, December 1977, 59. Images used to create ad: Woman with camera courtesy of Pro8mm. Modified sync recorder from *Super-8 Filmaker*, March/April 1974, 31. Super8 Sound Recorder and projector from *Super-8 Filmaker*, March/April 1974, 32. Edit Bench from *Super-8 Filmaker*, March/April 1974, 32. *Super-8 Filmaker* images courtesy of Paul Sheptow. Design by Joanne Chu.
50   Courtesy of Pro8mm.
51   Courtesy of Tommy Madsen.
52   Courtesy of Tommy Madsen.
53   Courtesy of Eastman Kodak Company.

CHAPTER 4

55   Photo by Ken Korsh. From *Super-8 Filmaker*, January/Feburary 1974, 34. Courtesy of Paul Sheptow. Design by Joanne Chu.
57   Collage by Victor Krummenacher. Kodachrome, Ektachrome, Tri-X boxes. Photos by Danny Plotnick. 4-X box. Photo by Alan Petty. Plus-X box. Photo by David Schendel.
Film Stills (L-R): Danny Plotnick's *Death Sled II: Steel Belted Romeos, Dream Syndicate, Pillow Talk, PIPSQUEAK PfOLLIES*, provided by Dino Everett, Archivist, USC SCA Hugh M. Hefner Moving Image Archive. *Drummer Needed* (group project produced at the University of San Francisco) provided by Movette Film Transfer.
58   Kodak Ad, "Kodachrome 40 Movie Film," 1975, Courtesy of Eastman Kodak Company.
59   Kodak Ad, "Kodak XL Movie Cameras," 1974. Courtesy of Eastman Kodak Company.
60   Photo by Danny Plotnick.
61   Kodak Ad, "Remember the Day in Pictures," 1968. Courtesy of Eastman Kodak Company.
62   Phil Vigeant shooting Max8. Courtesy of Pro8mm.
63   Courtesy of Pro8mm.
65   Courtesy of Pro8mm.

CHAPTER 5

All photos by Danny Plotnick, unless otherwise noted.
67   Bauer T2 Super, 1967–1968, $349.95.
72   Photo by Abigail Severance.
76   Unknown
77   Inflight photo: *Super-8 Filmaker*, March/April 1974, 28. Courtesy of Paul Sheptow.

CHAPTER 6

79, 81, 83 by Danny Plotnick.
79   Braun SB1 Viewer, $95 in 1975. Braun FK1 Splicer $34.95 in 1976.
85   Provided by Dino Everett, Archivist, USC SCA Hugh M. Hefner Moving Image Archive.

CHAPTER 7

89   "Sound Drive," 1973. From Kodak Historical Collection #003, D.319, Rare Books, Special Collections, and Preservation, River Campus Libraries, University of Rochester. Used with

	permission from Eastman Kodak Company.
90	Eumig ad, Mark S-810D, 1974.
91	Photos by Danny Plotnick.
92	Kodak ad, "Ektasound 240," 1977. Courtesy of Eastman Kodak Company.
93	From Kodak Historical Collection #003, D.319, Rare Books, Special Collections, and Preservation, River Campus Libraries, University of Rochester. Used with permission from Eastman Kodak Company.
94	Photos by Danny Plotnick.
95	Photo by Bette Duke. *Super-8 Filmaker*, July/August 1978, 20. Courtesy of Paul Sheptow.

CHAPTER 8

97	Kodak ad, "Kodak Consumer Centers," 1973. Courtesy of Eastman Kodak Company.
98	Photo by Danny Plotnick. Used with permission from Eastman Kodak Company.
99	Kodak ad, "Prepaid Processing," 1964. Courtesy of Eastman Kodak Company.
101	Photo by Danny Plotnick.
102	Photo by Florian Cramer. Courtesy of Super8 Reversal Lab.
104	Photo by Hein van Liempd. Courtesy of Super8 Reversal Lab.
105	Photo by Super8 Reversal Lab.

CHAPTER 9

107	Photo by Danny Plotnick.
109	Collage by Danny Plotnick. Images courtesy of Donald Deveau.
111	Photo by Danny Plotnick.

CHAPTER 10

113	Photo by Christian Bruno.
115	Film Chain at USC SCA Hugh M. Hefner Moving Image Archive. Photo by Danny Plotnick.
117	VHS and Beta SP photos by Christian Bruno. 3/4" & 1" photos by Danny Plotnick.
118	VHS design by Anthony Bedard & Danny Plotnick.
119	Kodak ad, "Supermatic Products" Courtesy of the George Eastman Museum. Used with permission from Eastman Kodak Company.

CHAPTER 11

121	Chinon DS-300, 1978, $575. Photo by Danny Plotnick.
123	Kodak Ad, "1973 Kodak Teenage Movie Awards Competition," 1973. Courtesy of Eastman Kodak Company.
124	Courtesy of Peggy Ahwesh.
127	Poster design by Silvana Nova. Courtesy of Marc Huestis.

CHAPTER 12

129	Photo by Danny Plotnick.
130	Photo by Danny Plotnick.
131	*Super-8 Filmaker*, March/April 1975, 26. Courtesy of Paul Sheptow.
132	*Super-8 Filmaker*, July/August, 1974, 47. Courtesy of Paul Sheptow.
133	*Super-8 Filmaker*, Winter 1972/1973, 11. Courtesy of Paul Sheptow.

INTERVIEWS

135	Kodak Ad, "Kodak Ektasound," 1973. Used with permission from Eastman Kodak Company.

INTERVIEWS 1960S/1970S

136	Kodak Ad, "The Instamatic Camera Idea to Movies," 1965. Courtesy of Eastman Kodak Company.
137	Kodak Ad, "The Instamatic Camera Idea to Movies," 1965. Courtesy of Eastman Kodak Company.

ROLAND ZAVADA

138	Photo by Danny Plotnick.
139	Kodak Ad, "You'll Find Us in Every Port," 1972. Courtesy of Eastman Kodak Company.
140	Kodak Ad, "Quad 8," Courtesy of the George Eastman Museum. Used with permission from Eastman Kodak Company.

LENNY LIPTON

141	Courtesy of Lenny Lipton.
142	Photo by Simon Bailey. *Super-8 Filmaker*, January/February, 1978, 45. Courtesy of Paul Sheptow.
144	Photo by Rod Wyatt. *Super-8 Filmaker*, December 1977, 32. Courtesy of Paul Sheptow.

JOHN PORTER

145	Photo by Edie Steiner. Courtesy of John Porter.
147	Photo by Blaine Spiegel. Courtesy of John Porter.

JAMES MACKAY

148	Photo by Liam Daniel © & courtesy Basilisk Communications Ltd.
151	*Journey to Avebury* © & courtesy LUMA Foundation. *Andrew Logan Kisses the Glitterati* © & courtesy LUMA Foundation. *In the Shadow of the Sun* © & courtesy LUMA Foundation. Collage by Danny Plotnick.
152	*B2 Movie* © & courtesy LUMA Foundation.

PAUL SHEPTOW

153	Courtesy of Paul Sheptow.
154	Collage by Danny Plotnick. Images courtesy of Paul Sheptow.

JONATHAN TYMAN

156	Courtesy of Jonathan Tyman.
159	Courtesy of Frank Uhle.

ROCKY SCHENK

160	Photo by Bill Paxton. Courtesy of Rocky Schenk.
161	Photos by Rocky Schenk. Collage by Danny Plotnick.
162	Photo by Carolyn Meltzer. Courtesy of Rocky Schenk.

JAMES NARES

164	Courtesy of James Nares.
167	Courtesy of James Nares.

BETH B

168	Courtesy of Beth B.
171	*Black Box* © 1979 Scott B and Beth B. Courtesy of Beth B.

NARCISA HIRSCH

172	Courtesy of Narcisa Hirsch.

INTERVIEWS 1980S/1990S

175	Courtesy of Norwood Cheek.

RICHARD LINKLATER

176	Courtesy of Detour Films.
177	Courtesy of Detour Films. Collage by Danny Plotnick

PEGGY AHWESH

180	Courtesy of Peggy Ahwesh.
181	Courtesy of Peggy Ahwesh.

DAVE MARKEY

183	Photo by Edward Colver.
184	Courtesy of Dave Markey.

G. B. JONES & BRUCE LABRUCE

187	Courtesy of Bruce LaBruce.
188	Photo by Candy Pauker. Courtesy of G. B. Jones.

MARTHA COLBURN

191	Photo by Martha Colburn.
193	Images by Martha Colburn. Collage by Danny Plotnick.

MATTHIAS MÜLLER

194	Courtesy of Matthias Müller.
195	Images courtesy of Matthias Müller. Collage by Danny Plotnick.

SILT
197 Photo by Jeff Warrin. Courtesy of silt.
198 Images courtesy of silt. Collage by Danny Plotnick.

NORWOOD CHEEK
202 Courtesy of Norwood Cheek.
203 Courtesy of Norwood Cheek.

MELINDA STONE
204 Courtesy of Melinda Stone.
205 Photos by Melinda Stone.

INTERVIEW 2000S
208 *Super-8 Filmaker*, March/April 1978, 20. Courtesy of Paul Sheptow.

LISA MARR & PAOLO DAVANZO
209 Photo by Naomi Uman. Courtesy Marr/Davanzo.
211 Courtesy Marr/Davanzo.
212 Courtesy Marr/Davanzo.

ED SAYERS
213 Photo by Alex Glynn and Team Straight 8.

KARISSA HAHN
214 Courtesy of Karissa Hahn

LEGACY
217 From 1965 Kodak brochure, "This Is the New Dollar Sign in Movies…" Courtesy of Eastman Kodak Company.
218 Courtesy of Harvard Film Archive.
219 Photos © Les Levine 1972.
220 Photo © Melinda Stone and Igor Vamos. Courtesy of Melinda Stone.
221 Courtesy of David Beydler and the Academy of Motion Picture Arts and Sciences Film Archive.
223 Courtesy of Georg Marioth.
224 Courtesy of Bill Daniel.
225 Photo by Bruce Anderson, from *Super-8 Filmaker*, 30, March/April 1978. Courtesy of Paul Sheptow.

INDEX

1" video, 28, 95, 116, 117, 118, 218
16mm
  history, 14, 16, 75, 77, 138, 143, 149, 156, 157, 192, 220
  reduction printing to Super 8, 16, 111, 141
  super 8 printing to 16, 40, 103, 110, 144, 152, 155, 179, 196
  vs. Super 8, 10, 14, 15, 16, 18, 20, 25, 32, 40, 48, 56, 59, 61, 64, 76, 80, 83, 102, 136, 137, 141, 142, 143, 145, 146, 156, 163, 172, 180, 183, 192, 194, 199, 215, 225
  workflow, 40, 44, 58, 60, 68, 70, 72, 100, 101, 108, 143
18 vs. 24 fps, 16, 40, 82, 114, 136
3/4" Video (U-matic video), 24, 46, 116, 117, 122, 125, 152, 166, 186, 218
35mm, 14, 64, 72, 100
  cutting 35 down to Super 8, 60, 61, 62, 63, 64, 104, 141
  reduction prints, 111
  Super 8 prints on 35, 16, 139
  vs. Super 8, 10, 25, 40, 48, 56, 61, 145, 149, 158, 181, 192, 200, 215
3-D Super 8, 142, 144, 147
3M, 56
4-X, 57, 59, 60, 130, 141, 152, 162, 175
40 Watt Club, 203
911 Media Arts Center, 122
AEIOU Guide, 206
Abrams, J.J., 16
Academy of Media Arts (KHM), 194
Adlestein, Gary, 221
ADOX, 61, 104, 105
Agfa, 34
Agfa film stock, 56, 61, 104, 105
Agfa CHS II, 104
Agfa Microflex Sensor, 43
Agfa Moviechrome, 102
Agfa Movexoom 3000, 43
Agfa Movexoom 4000, 11, 19
Ahearn, Charlie, 20, 167
Ahearn, John, 169
Ahwesh, Peggy, 174, 180–183
  The Color of Love, 182
  The Deadman, 183
  Martina's Playhouse, 181, 183
  Philosophy in the Bedroom, 180
  Pittsburgh Trilogy, 180

Akerman, Chantal, 178, 214
Alte Kinder Film Collective (see Müller, Matthias)
Amateur 8 Movie Contest, 162
American Cinematographer, 48, 155
AMP Sound Super 8 Model 2, 41
Anarchistische Gummizelle, 196
Andec, 103, 104
Andersen, Yvonne, 156
Anderson, Keith, 108
Anderson, Wes, 20
Anger, Kenneth, 165
Ann Arbor 8mm Film Festival, 28, 122, 126, 137, 147, 156–160, 162
Ann Arbor Film Festival, 156–158, 214
Anthology Film Archive, 166, 167, 214
Antioch College, 180
Argus, 16
Arnheim, Rudolf, 159
Artists Space, 166, 168, 182
Artists' Television Access 122, 126
Athens International Film Festival, 122
Attack of the 50 Foot Reels, The, 204
audio cassette (see cassette tape)
automatic exposure (AE) lock, 34, 40
Azar, Caroline, 188, 189
B, Beth, 20, 136, 164, 168–171, 182
  Black Box, 168, 169, 171
  Exposed, 171
  G-Man, 168, 169
  Letters To Dad, 168, 169
  The Offenders, 169, 170
  Salvation!, 168
  Two Small Bodies, 168
B, Scott, 20, 136, 164, 168, 170, 182
back light control (BLC), 34
Bags, The, 186
Baille, Bruce, 142
balance stripe (see sound stripe)
Ballard, J. G., 188
B & T's Little Film Notebook (see Brodsky & Treadway)
Bangs, Lance, 203
Bard College, 180
Barnes & Barnes, 160, 163, 164
barney, 90
Bauer, 16, 62, 144
Bauer C14 XL, 43
Bauer T2 Super, 67
Bazin, André, 178

Beards, The, 210
Beaulieu, 24, 32, 34, 40, 44, 62, 64, 141, 150, 155, 158, 164, 165
Beaulieu 4008 ZMII, 41, 141
Beaulieu 7008, 186
Beaulieu, Marcel, 32
Bedard, Anthony, 118
Beebe, Roger, 204
Bell, Bob, 199
Bell & Howell, 14, 16, 20, 44, 139, 142, 158, 202, 210
Bell & Howell Autoload 374F, 31
Bell & Howell Filmosonic 1227 XL, 42
Bell & Howell Filmosound 8, 44, 48, 49
Benning, James, 178, 214
Benson, Greg, 220
Benzine, Craig (*How I Feel After 11 Years on YouTube*), 218
Bergman, Ingmar, 187
Berlin Super 80, 221, 222, 223
Beta SP, 116, 117, 192, 218
Beths, The, 203
Beydler, Gary (*Mirror*), 221, 222
Black Flag, 183, 185
Blockbuster Video, 186
Blum, Chris, 153
Bokanowski, Patrick, 214
Bolex, 82, 144, 149, 150, 152, 172, 194, 202
Bolex 18-5L, 8
Bolex 160 Macrozoom, 37
Bolex 280 Macrozoom, 43
Bolex Striping Machine N8/S8, 91
Boone, David (*The Invasion of the Aluminum People*), 176
Borchardt, Mark (*American Movie, The More the Scarier*), 225
Boston Underground Film Fest, 122
Boyle, Bern, 127
Brakhage, Stan, 14, 149, 174
Braun, 16, 24, 32
Braun FK1 Splicer, 79
Braun SB1 Viewer, 79
British Film Institute, 152
Brodsky, Bob (see Brodsky & Treadway)
Brodsky & Treadway, 70, 192
  B&T's Little Film Notebook, 204
  Super 8 in the Video Age, 90
Bromberg, Betzy, 214

Brooks, Mel (*High Anxiety, Young Frankenstein*), 184, 185
Broughton, Richard, 142
Broughton, James, 142
Bruinsma, Frank, 102–105
Brundert, Dagie, 101
Brüning, Jürgen, 190
Brussels Super 8 Film & Video Festival, 122
c-mount lenses, 32
Cadena, Dez, 185
California Institute for the Arts, 214
Caldini, Claudio, 18
camera original, 60, 68, 70, 72, 94, 95, 108, 110, 114, 118, 125, 182, 218
cameras, 30–53
  functions, 14, 16, 20, 28, 32, 34, 40, 44, 52, 58, 64, 65, 99, 100, 144, 147, 150, 162, 167
  history, 10, 16, 20, 32, 34, 44, 48–53, 58, 63, 101, 130, 131, 137, 139–141
  sound recording, 18, 32, 34, 44, 48–50, 75, 82, 90–93
  portability, 32, 145, 148, 165, 166, 169, 179, 181, 220, 221
Cannes Film Festival, 213
Canon, 16, 32, 34, 44, 176, 184
Canon 310, 209
Canon 310 XL, 65
Canon 312 XL-S, 19
Canon 514 XL-S, 34, 40
Canon 1014 XL-S, 32, 35, 40, 192
Canon Auto Zoom 1014, 42
Canon Auto Zoom 1218, 39
Canyon Cinema, 142
Caracas Super 8 Festival (see International Festival of the New Super-8 Cinema)
Carrie, 163
cartridge, 20, 44, 50, 52, 63, 64, 98, 100, 101, 104, 119, 138–141, 165, 181, 197, 204, 207, 213
  50-foot, 14, 32, 40, 101, 130
  200-foot, 40, 143
  reloadable, 63, 64
  sound, 139, 140, 143
Cassavetes, John, 14
Castillo, Carlos, 159
cassette deck, 18, 27, 83, 149, 162, 192
cassette recorder, 18, 44, 48, 49, 162, 165, 214

cassette tape, 18, 27, 48, 85, 127, 150, 158, 162, 188, 190, 192
Castro Camera, 126
Castro Theater, 61, 72
CBGB, 169
Consumer Electronics Show, 52
Chandler, Jasper, 138, 139
Cheek, Norwood (also see *Flicker*), 174, 202-204
    Mower, 203
    Sex Police, 202
Chicago Underground Film Festival, 122
Chinon, 34, 40, 62, 64
Chinon 506 SM XL Direct Sound, 36
Chinon 877 Macro, 8
Chinon DS-300, 121
Chinon Pacific 200, 203
Chinon SP 330 MV, 24, 26, 70
Chladek, Jim, 166
Cine Gear, 51
Cine Slave, 49
Cine-Action!, 188
CineCycle, 182
Cinédia, 103
Cinema of Transgression, 20
Circle Jerks, The, 185
Club 57, 165, 170
Colburn, Martha, 174, 191-193, 207
Cole, Robert, 48
Collective for Living Cinema, 182
Comden, Tippi, 183
Conrad, Tony, 165, 180
Corman, Roger, 16
Cortez, Diego, 167
Cosina, 40
Cronenberg, David, 182
CROSSROADS, 214
Cub, 210
Cut-a-Rut, 91, 153
Cyzyk, Skizz, 192
Danhier, Celine (*Blank City*), 165
Daniel, Bill (*The History of Texas City*), 206, 222, 224
Daniel, Lee, 176, 178, 179
Davanzo, Paolo, 101, 207–212 (also see Echo Park Film Center)
    Freetime and Sunshine, 210
    The Sound We See, 211
De Grasse, Herb, 136, 137, 142
Fialka, Gerry, 137
DeGeneres, Vance, 136
Deren, Maya, 136, 167, 174, 182
Dick, Vivienne (*Staten Island*), 20, 164, 167, 170, 222
Die Tödliche Doris, 194
Dinosaur Jr., 184
dissolves, 32, 40, 162, 203
double Super 8, 44, 144
double system sound (see sound)
Doyle, Bob, 18, 44, 48–50
Duggan, Dennis, 95, 153, 155
Dwayne's Photo Lab, 58
Easy TV, 186
Echo Park Film Center, 101, 208–212
edge numbers, 68, 70
Edison, Tommy, 218
editing, 18, 44, 49, 61, 68, 78–87, 92, 144, 146, 147, 149, 158, 160, 162, 182, 183, 185, 190, 201, 213, (also see sound stripe, splicers, sound-on-sound, splicers, splices, viewers)
editing bench, 44, 144
editors (see viewers)
Edwards, Evan, 138, 139
Eisenstein, Sergei (*Strike*), 24, 25
Eiszeitkino, 197
Ektachrome, 52, 58, 59, 60, 63, 64, 100, 102, 103, 105, 108, 130, 152, 167, 170, 200
Ektachrome 64T, 58, 59
Ektachrome 100D, 58, 59, 208
Ektachrome 125, 58, 59
Ektachrome 160, 58
Ektachrome 160 Type A, 58, 59, 60, 130, 155, 156
Ektacrhome 160 Type G, 57, 58, 59, 130, 200
Electronic Eye (EE) Lock, 34, 40
Elmo, 34, 40, 77, 82, 157, 176, 181, 184, 201
Elmo 1012S-XL Macro, 43
Elmo GS 1200, 74, 75, 181
Elmo GS 1200 Xenon, 75
Elmo S1 180E, 185
Elmo ST 1200D, 75, 181
Elmo Super 106, 13
Epstein, Rob, 127
Eumig, 34, 44, 62, 143, 199, 209
Eumig Mark 610 D, 68
Eumig Nautica, 203
Evans, Keith (see silt)
experi & nixperi, 197
exposure, 28, 34, 40, 60, 65, 85, 99, 100
    for prints and transfers, 110, 114, 116
f-stop, 28, 34, 100
fades, 40, 162
Farrell, Christian (see silt)
Fellini, Federico, 186
Ferrania, 61, 105
Fifth Column (see Jones, G. B, and Porter, John)
Film Arts Foundation, 122, 207
film chamber, 14, 34
Film Culture, 187
Film Forum, 170
film gate, 14, 40, 50, 52, 65, 87, 90, 92, 138–140, 149, 204
film labs, 10, 14, 44, 58, 64, 68, 70, 92, 94, 95, 98–100, 102–105, 108, 110, 114, 139, 143, 158, 164, 180, 190, 204, 211, 212
film stock, 10, 14, 18, 20, 40, 54–65, 75, 82, 83, 92, 98–101, 104, 105, 108, 110, 130, 148, 162, 167, 175, 180, 192, 195, 200, 206, 208 (also see 3M, 4-X, ADOX, Agfa, Agfa CHS II, AGFA Moviechrome, Ektachrome, Ektachrome 64T, Ektachrome 100D, Ektachrome 125, Ektachrome 160, Ektachrome 160 Type A, Ektacrhome 160 Type G, Ferrania, Fuji, GAF, Kodachrome, Kodachrome II, Kodachrome 40, Orwo, Tri-X, Wittner, print stocks)
Films and Filming, 187
filters (85A, ND), 40, 58
filter pack, 95, 108
Firehose, 184, 186
Fisk, Jack, 163
Fitzgibbon, Coleen, 170
Five-Eight, 203
fixed mount lenses, 34
flatbed, 18, 44, 49
Flicker screening series, 175, 202, 203, 204
Flicker zine, 175, 202, 204
flying spot scanner, 50, 114, 119
Fountain, Ken, 212
Frameline Film Festival, 126, 127
Freed, Hermine, 165
French New Wave, 20
Frost, Jeff (*Flawed Symmetry of Prediction*), 222
Fuji, 44
    film Stock, 56, 200
    Single 8, 44, 45, 77, 139, 140
    processing, 99
Fujica AX100, 45
Fujica Marine-8, 45
Fujica Single 8 P400, 45, 77
Fujica Single 8 Z600, 43
Fujicascope SH10, 77
fullcoat, 44, 48, 49, 63
Funnel, The, 122, 145, 146, 147, 148, 182, 187, 188, 189, 190
GAF, 16, 46, 56
GAF 805 M, 25, 46, 47
Gallagher, Stephen, 182
Galluzzo, Tony, 153
Gauge Film, 103
Gay Film Festival of Super-8, The, 126–127
Gegenlicht, 18
Gesualdi, Vito (*Berkeley Protesters Take the Pepsi Challenge*), 220
Gibbons, Joe, 182
Gibney, Alex (*Going Clear, Taxi to the Darkside*), 20, 137, 156
Gidal, Peter, 149
Gilliam, Terry, 191
Gilligan, Vince, 16
Girardet, Christoph, 196
Global Super 8 Day, 122
Godard, Jean Luc, 14, 187
Goethe Institute, 195
GOKO, 82, 181 (also see sound recording in postproduction and sound stripe)
    GOKO RM-5005, 82, 86
    GOKO RM-8008, 82, 83, 86, 94, 95
Gonzales, David, 127
Gonzales, Greg, 127
Gorman, Don, 58
Grass, Angie, 203, 204
Gravelle, David, 189
Green, Hank, 218
Green, John, 218
Greendale, (see Young, Neil)
Haghefilm, 104
Hahn, Karissa (*Please step out of the frame*), 208, 214–215
Hallwalls, 182, 190
Hama, 87
Hamton Engineering, 49
hand processing, 72, 100–104, 181, 182, 194, 201, 207, 210, 212
Hanoi Doc Lab, 211
Hansen, Brian (*Speed of Light*), 176
Happenings, 157, 172
Harvard, 18, 44, 48, 49
Haynes, Todd (*Carol, Safe, The Suicide*), 20, 137, 156
Hervic splicer, 81, 87
Hervic splices 81
Heurtier Stereo 42, 71
Heuwinkel, Christiane, 194, 195
Hilsenrad, Joe, 212

Hindle, Will, 142
Hinrichs, Kit, 153
Hirsch, Narcisa, 18, 137, 172-173
   Bebes, 173
   Come Out and Show Them, 172, 173
   La Marabunta, 172
   Pink Freud, 173
   Portraits, 173
Hitchcock, Alfred, 111, 176
Hollywood Home Video
home movie, 186
Homocore (see Queercore)
Honolulu Underground Film Festival, 122, 174
Houston International Film Festival, 176
Howland, Don, 186
Huestis, Marc, 126, 127, 156
Huillet, Danièle, 149
Hurrah, 169
Impossible Project, 65
Inflight Motion Pictures, 77
International Festival of the New Super-8 Cinema, 126, 155, 158
intervalometer, 32, 40
Ism, Ism, Ism: Experimental Cinema in Latin America, 172
Jarman, Derek, 18, 116, 117, 137, 148-152, 191, 194
   Andrew Logan Kisses the Glitterati, 151
   Angelic Conversation, 150
   Dancing Ledge, 194
   Derek Jarman Super 8, 149, 194
   Garden, The, 150
   Gerald's Film, 150
   Imagining October, 150
   In the Shadow of the Sun, 150, 151, 152
   Journey to Avebury, 150, 151
   Jubilee, 148, 150
   Last of England, The, 137, 150
   Miss Gabby Gets it Together, 150
   Picnic at Rae's, 150
   Sebastiane, 137, 148, 150
   Studio Bankside, 150
   Sulphur, 150
   Tarot, 150
   Tempest, The, 150
Jarmusch, Jim, 10, 164, 174, 178, 199,
   Stranger Than Paradise, 199
   Year of the Horse, 61
Jelinski, Manfred, 182
Jensen, Jytte, 221

Johnston, Becky, 164, 166, 170
Jones, G. B., 145, 174, 187-191
   Don't Be Gay or How to Stop Worrying and Fuck Punk Up the Ass, 190, 191
   Fifth Column, 145, 187-189, 191
   Hide, 188
   J.D.s, 187-189
   Lollipop Generation, The, 191
   Second Unit, 188, 189
   Troublemakers, The, 188, 189
   Unionville, 188
   Yo-Yo Gang, The, 189, 191
Jordan, Larry, 142, 174, 191
Jost, Jon, 178
Kahl, 61, 105
Kanai, Catherine, 221
Kern, Richard, 20, 170, 186
Kiefer, Anselm, 200
Kin-O-Lux, 182
Kinetta 5K Scanner, 117
Klahr, Lewis, 174, 182, 191
Kober & Döbele, 196
Kodachrome, 56, 58, 59, 63, 64, 102, 108, 141, 144, 167, 200
Kodachrome II, 56, 138, 141
Kodachrome 40, 27, 29, 56, 57, 58, 59, 130, 152, 180
Kodak Brownie, 15, 184
Kodak Ektasound 130, 18, 93
Kodak Ektasound 140, 18, 93
Kodak Ektasound 160, 39
Kodak Ektasound 240, 33, 90, 92
Kodak Instamatic M109 K, 71
Kodak labs & processing, 58, 98-100
Kodak M2, 41
Kodak M4, 21
Kodak M95, 71
Kodak Moviedeck, 70, 199
Kodak Moviedeck 457, 71
Kodak negative stocks, 61, 64, 208
Kodak new Super 8 camera, 10, 50-53, 131
Kodak Picture Spot, 204, 220
Kodak Pre-Paid Mailer, 98, 99
Kodak Presstape, 83, 84, 87
Kodak–Super 8 history, 10, 14, 20, 44, 50, 52, 62, 75, 90, 131, 138-140, 142, 196
Kodak Supermatic 8 Processor, 101
Kodak Supermatic 200, 40
Kodak Supermatic Video Player, 50, 119
Kodak Universal Splicer, 83, 84, 87
Kodak XL 33, 15

Kodak XL 320, 38
Kodak XL cameras, 58, 59, 139
Korvette Convertible 8ZR, 8
Kitchen, The, 124, 182
Kuchar, George, 181, 187, 191
Kuchar, Mike, 181, 187, 191
Kuhlbrodt, Dietrich, 197
Kunsthalle Bielefeld, 194
L'Hotsky, Tina, 166
LA Reader, The, 186
LA Weekly, The, 186
LaBruce, Bruce, 174, 187-191
   Boy, Girl, 188, 189
   Bruce and Pepper Wayne Gacy's Home Movies, 188, 189
   Don't Be Gay or How to Stop Worrying and Fuck Punk Up the Ass, 190, 191
   I Know What It's Like to Be Dead, 188
   J.D.s, 187-189
   No Skin Off My Ass, 188, 190
   Slam, 188
   Super 8 ½, 187, 190
Land, Edwin, 48
Le Grice, Malcolm, 149
Leacock, Richard, 18, 44, 48, 49
Lee, Craig, 186
Lee, Spike, 20
Levine, Les (*The Troubles: An Artist's Document of Ulster*), 219, 221
Levine, Saul (*Raps and Chants, Time to Go to Work*), 221
Lhasa Club, The, 185, 186
Lidstone, John, 152
Light Cone, 103
Light Industry, 183
light meter, 20, 28, 32, 34, 100
Linklater, Richard, 20, 174, 176-180
   Austin Film Society, 176
   Boyhood, 20, 176
   Dazed and Confused, 20, 174, 176, 179
   It's Impossible to Learn to Plow by Reading Books, 176, 178, 179
   School of Rock, 176
   Slacker, 20, 174, 176, 177, 178, 179
Liotta, Jeanne, 195
Lipman, Ross, 110
Lipton, Lenny, 18, 32, 131, 136, 141-145, 147, 153, 220 (also see *3-D Super 8*, *The Super 8 Book*)
   Children of the Golden West, 136, 137, 141
   Uncle Bill and the Dredge Dwellers,

145
Lipzin, Janis Crystal, 174, 180
Local 506, The, 203
Logmar Camera Solutions (see Madsen, Tommy)
London Filmmakers Cooperative, 122, 149, 182
Lounge Lizards, The, 170
Loud, Lance, 164
Lunch, Lydia, 164, 168, 170, 171
Lurie, John, 20, 164, 165
Mackay, James, 117, 148-152
Madsen, Tommy, 50-52, 131
magnetic stripe (see sound stripe)
Magoo's, 169
Manhattan Cable, 166
Mann, Ron, 147
Manupelli, George, 157
Markey, Dave, 100, 116, 136, 174, 183-187
   1991: The Year Punk Broke, 184, 186, 187
   Desperate Teenage Lovedolls, 183-186
   Devil's Exorcist, The, 184
   Lovedolls Superstar, 186
   Movie of Movies, The, 136, 184
   Omenous, The, 184, 185
   Painted Willie, 184
   Sin 34, 184
   Slog, 184, 185
   We Got Power, 184
Maron, Marc, 218
Marr, Lisa, 101, 208-212 (also see Echo Park Film Center)
   Freetime and Sunshine, 210
   The Sound We See, 211
Mars, 170
Maslin, Janet, 170
Matzkin, Mike, 153
Max's, 169
Max 8, 65
Maximumrocknroll, 190
Maybury, John, 149
McClard, Michael, 166
McLaren, Norman, 148, 204
Mclaren, Ross (*Crash and Burn*), 187
Mears, Ric, 127
Media Arts Centers, 20, 116, 122, 123, 180, 182, 204, 206, 209-212 (also see 911, Echo Park Film Center, Film Arts Foundation, London Filmmakers, Pittsburgh Filmmakers, Squeaky Wheel)
Megaphone Records, 192

Mercer Media, 167
Merge Records, 202
Miami Underground Film Festival, 122
Micro Cine Fest, 122
Mid Atlantic Skate & Sound Festival, 29
Miggens, Billy, 127
Millennium Film Workshop, 166, 167, 182
Minette, 83
Minette S-5 Viewer, 81
mini-input, 32, 34, 83, 90
MiniDV, 28, 11
Minolta, 34
Minolta Autopak-8 D 10, 41
Minolta Autopak-8 D 6, 43
Minolta Autopak-8 K11, 8
Minolta XL 660 Sound, 43
MIT, 18, 44, 48, 49
Milk, Harvey, 126, 127
Mitchell, Eric (*Bikers, Car Crash*), 164, 165
Modern Photography, 145, 153, 155
Montreal Festival du Jeune Cinema, 122
Montreal Gay and Lesbian Film Festival, 189
Moore, Thurston, 184
Morrissey, Paul, 165
Mosset, Olivier, 165
Moving Image, 153
Mr. Bill, 136
MTV (*120 Minutes*), 163, 178, 203
Mudhoney, 184
Muffs, The, 184, 210
Müller, Matthias, 174, 194-197, 200
    Alte Kinder, 194, 196, 197, 199
    Aus der Ferne–The Memo Book, 195, 196
    Final Cut, 196
    Pensão Globo, 195
    Sleepy Haven, 196
multi-track mix, 83, 85, 86, 94, 114
Nagra, 146, 210
Nares, James, 20, 116, 137, 164-167, 220
    Arm & Hammer, 164
    Block, 164
    The Contortions, 164, 167
    No Japs at My Funeral, 164, 167
    Pendulum, 166
    Ramp, 220
    Rome '78, 164, 166, 167
    Steel Rod, 164
    Street, 166

Waiting for the Wind, 164, 166, 167
National Film Board of Canada, 145, 146
Necro Films, 225
negative film stock, 60, 61, 63, 64, 65, 68, 70, 104, 108, 110, 130, 131, 208
Neistadt, Casey (*Bike Lanes*), 220
Nelson, John, 156
Nelson, Robert, 142
Neri, Julio (*Erase Una Vezen Venezuela*), 220, 225
New Cinema, The, 166, 167
New York Underground Film Festival, 29, 122
Nguyen, Thi, 211
Nicoletta, Danny, 126, 127
Nikon, 16, 32, 153
Nikon R 10, 32, 43
Nirvana, 184
Nizo, 18, 32, 40, 62, 90, 144, 145, 150, 165, 167, 209
Nizo 801, 39
Nizo 6080, 63
Nizo Integral, 63
Nizo Integral 7, 42
Nizo S 30, 197
Nizo S 560, 40, 147
No Wave, 20, 164, 168, 170, 174
North East London Polytechnic, 149
O'Toole, Owen, 195, 196
Oberhausen International Short Film Festival, The, 197
Once, The, 157
Onion City Film Festival, 122
Ontario College of Art & Design (OCAD), 187, 188
Optasound, 49
Orwo, 61, 104, 105
Osborne, Steve, 111
Osnabrück Experimentalfilm Workshop, 197
Otterness, Tom, 170
Padeluun, 194
Page, Bettie, 191
Paillard, 143
Parish, James, 204
Pasolini, Pier Paolo, 187
Pathé, 44
Pauker, Candy (*Dr. Smith, Interview with a Zombie*), 188, 189
Paxton, Bill, 160-163
Pearce, Terry, 90
Peck, Ron (*What Can I Do With a Male Nude?*), 152
Pennebaker, D. A., 14

perforations (perfs), 44, 63, 64 82, 138, 143
pin-registered Super 8 camera, 50
Pincus, Edward (*Guide to Filmmaking*), 184
Pittsburgh Filmmakers Co-op, 122, 180, 182, 183
Pixelvision, 183
Place, Pat, 164, 167
platen, 82
Plotnick, Danny, 24, 25, 46, 76, 77, 84, 125, 126
    Death Sled II: Steel Belted Romeos, 28, 57, 124, 125
    Dream Syndicate, 57
    Dumbass From Dundas, 124
    I'm Not Fascinating–The Movie!, 46, 118, 119
    Pillow Talk, 57
    PIPSQUEAK PfOLLIES, 57, 94, 95, 118
    Skate Witches, 27-29, 84, 85, 124
Plus-X, 57, 59, 94, 130, 138, 152, 180
Poe, Amos, 20, 164
Polaroid, 48, 65
Polavision, 101
Polke, Sigmar, 200
Popular Photography, 141, 142, 143, 145, 155
portapak, 24, 165, 167, 209
Porter, John, 136, 137, 145-148, 187, 189
    Centrifuge, 148
    Drive-In Movie, 147, 148
    Porter's Condensed Rituals, 145, 147, 148
    Santa Claus Parade, 145, 147, 148
    Scanning, 148
post-stripe (see sound stripe)
Powell, Michael (*Peeping Tom*), 196
pre-striped stocks (see sound stripe)
pressure pad/plate, 44, 52, 138, 139
prints, 68, 70, 94, 95, 103, 106-111, 114, 124, 125, 139, 141, 143, 144, 148, 158, 182, 183
    contact printing/prints, 108, 110
    one-light prints, 110
    optical printing/prints, 108, 110, 79, 182
    print stock, 95, 103, 108, 110, 152
    print stock 7272, 110
    print stock 7361, 108
    print stock 7385, 110
    print stock 7399, 103, 108, 110
    reduction prints 16, 111, 139, 144, 183

sound prints, 94-95, 110, 139
timed prints, 110
Pro8mm (see Phil Vigeant)
processing, 58, 65, 96-105 141, 149, 187 (also see hand processing)
projectors, 10, 24, 25, 40, 44, 52, 66-77, 92, 114, 116, 117, 130, 138, 139, 141, 143, 144, 148, 149, 150, 157, 159, 183, 185, 186, 196, 199, 200, 201, 212, 215, 224, 225
    for postproduction sound, 18, 27, 48, 75, 83, 85, 86, 90, 165, 192
    throw, 73, 75, 76, 196
    cartridge-loaded projector, 77
projector bulbs, 72, 76, 77, 196
Pryce, Charlotte, 214
Pylon, 203
Queercore, 145, 187, 189, 190, 191
Raimi, Sam, 16, 136
Ramones, 169, 184, 185, 190
Rams, Dieter, 32, 197
Reble, Jürgen, 201
Redd Kross, 183, 185
reduction prints (see prints)
Reel Image, The, 111
registration pins, 82, 84
Regular 8mm, 14, 51, 156, 182, 184, 199, 210
    history, 14, 138, 140
    vs. Super 8, 14, 15, 16, 146, 225
Reich, Steve, 172
reversal vs. negative workflow, 68, 80, 108, 110
reversal film stock, 52, 56-61, 63, 80, 100, 103, 104, 105, 148, 214
Rhodes, Lis, 149
Rhonda Camera, 65
Richter, Chris, 52
Richter, Suzie, 189
Rimmer, David, 200
Ritz Camera, 24
Robertson, Anne (*Five Year Diary, Apologies*), 218, 220
Russell, Ken (*The Devils, Savage Messiah*), 148, 149, 150
Ryan, Cory, 204
Ryerson Polytechnical, 145, 146
San Francisco State University, 174, 197, 199
Sankyo, 16, 40, 64
Saturday Night Live, 136, 162, 184
Sawyer, Paul, 142
Sayers, Ed, 213
Sayles, John (*Return of the Seacaucus Seven*), 176

Schaefer, Dirk, 196
Schenk, Rocky, 136, 137, 156, 160-164
   Dream Sequence, 160, 162
   Egyptian Princess, The, 160-164
   Fish Heads, 160, 163
   Killer Chihuahuas, 162
Schmidt, Martha, 158
Schofill, John, 142
School of Visual Arts, The, 164, 168, 169
Schmelzdahin (*Stadt In Flammen, Wir lagerten wie gewöhnlich um's Feuer, Weltenempfänger, Der General, The Flamethrowers*), 194, 195, 196, 197, 199, 200, 201
Schroeter, Werner, 196
Seery, Bill, 167
Sharits, Paul, 149, 174, 180
Shattuck, Kim, 210
Sheptow, Paul, 153-156 (also see *Super-8 Filmaker*)
Sherman, Cindy, 166
Shonen Knife, 184, 186
Sikora, Jim, 125, 174
silt, 174, 197-202
   Conflict, 200
   Kuch Nai, 200, 201
Single 8, 44, 45, 77, 139, 140, 200
single frame, 40, 44, 141, 147, 202
single system sound (see sound)
Sirko, Radek (*Poglos Reverberation*), 222
Smith, Jack, 181, 187, 191
Smith, John, 149
Smolen, Wayne, 127
Snow, Michael, 165, 174
Sonic Youth, 184, 186
Sony TC-800B, 48
sound, 88-95
   double system sound, 18, 44, 48-50, 137, 143, 144, 156, 159, 165 (also see Bell & Howell Filmosound, cassette deck, cassette recorder, cassette tape, Bob Doyle, Hamton Engineering, Richard Leacock, MIT, Sony TC-800B, Phil Vigeant,)
   single system sound, 18, 32, 44, 90, 136, 137, 140, 141, 143, 150, 158, 165, 187
   sound cameras, 32, 34, 52, 181
   sound head, 34, 70, 82, 85, 91, 92, 118, 140
   sound-on-sound, 82, 85, 86, 185
   sound projectors, 75, 76, 83
   sound recording in postproduction, 27, 44, 75, 82, 83, 85, 86, 94, 95, 162, 185, 190, 192
   sound stocks, 10, 75, 82, 83, 130, 175, 192
   sound viewers, 82, 83, 85, 86
   sync sound 18, 20, 32, 118, 130 (also see single or double system)
   two-track sound, 75, 86, 87, 185
sound stripe, 18, 44, 75, 84, 85, 87, 90, 91, 110, 130, 143, 153, 162
   main stripe 82, 83, 85, 86, 87, 90, 94, 185
   balance stripe, 75, 82, 83, 85, 86, 87, 94, 165
   post-stripe, 92, 94, 95, 110, 183
Splice This!, 122
splicers, 80, 82, 84, 87, 146
splices, 80, 84, 87, 127, 146, 167, 192
   film cement/cement splices, 80, 149, 162
   tape splices/splicing tape, 44, 80, 84, 87, 149, 157, 162, 174
Squeaky Wheel Media Arts Center, 122
Standard 8mm (see Regular 8mm)
Standardization of Super 8, The, 139
Stark, Scott (*11/9/85/Las Vegas/NV*), 220
Steenbeck, 49
Stone, Melinda, 123, 175, 204-207, 220
   Barbie Liberation Organization, 204, 206
   Center for Land Use Interpretation Photo Spot Project, 204, 220
   Fleur Power, 206
   The California Tour, 205
Stone, Oliver (*Natural Born Killers*), 61
stop-motion animation, 174, 191, 192, 202
Straight 8 Film Festival, 208, 213
Straub, Jean-Marie, 149
Studio Één (Studio One), 102, 103, 105
Suicide, 166
Sundance Film Festival, 152, 197, 201
Sunset Sound, 186
Suntar 303 Dual 8 Viewer, 83
Super 8 Book, The, 141, 142
Super 8 in the Video Age (see Brodsky & Treadway)
Super 8 Motel, 124, 182
Super 8 Solar System, 182
Super Super 8 Festival, 122, 123, 175, 204-207 (also see Stone, Melinda)
Super-8 Filmaker, 137, 142, 143, 147, 153-156, 160 (also see Sheptow, Paul)
Super8 Reversal Lab, 102-105 (also see Bruinsma, Frank)
Super8 Sound, 18, 48-50, 61, 62-65, 155 (also see Doyle, Bob and Vigeant, Phil)
Superchunk, 202, 203
Supersound Film Striper, 91
sync sound (see sound)
synchronizer, 49, 144
Takita printer, 103
tape deck (see cassette deck)
tape recorder (see cassette recorder)
Tati, Jacques, 214
Taubin, Amy, 165, 167
Teague, David, 204
Technicolor, 63, 64
Technicolor Mark Ten, 43
Teenage Jesus and the Jerks, 170
tENTATIVELY a cONVENIENCE, 192
time lapse, 32, 40, 203, 204, 222
Trans Com, 76
Treadway, Toni (see Brodsky & Treadway)
Tri-X, 57, 59, 60, 64, 103, 105, 130, 138, 148, 152, 162, 203, 214
TSOL, 185
tungsten light, 40, 58, 59, 63
Tyman, Jonathan, 156-159
Uher, 210
underwater cameras, 44
University of California, San Diego, 206
University of Michigan, 24, 84, 156-159, 174
University of San Francisco, 159, 206
University of Texas, 178
US Super 8 Film & Video Festival, 122
Vamos, Igor, 204, 206, 220
Van Buren, Richard, 164
Van Meter, Ben, 142
van Oosten, Marsel (*Namibian Nights*), 222
Varda, Agnes, 187
variable shutter, 32, 40
VCR, 25, 123
Vega, Alan, 166
Vega, Arturo, 169
Versum, Uli, 194
VHS, 28, 29, 62, 94, 116, 117, 124, 125, 156, 157, 179, 186, 190, 206, 209
video & digital transfer, 27, 28, 60, 61, 65, 68, 94, 95, 108, 112-119, 122, 125, 143, 148, 152, 166, 184, 186, 192, 218
viewers, 68, 78-86, 90, 101, 183, 210
Vigeant, Phil, 51, 52, 62-65
Village Voice, The, 186
VIVA 8, 122
Von Praunheim, Rosa, 152
Vuntut Gwitchin Nation, 211
Wang, Wayne (*Chan is Missing*), 176
Warhol, Andy, 165, 166, 173, 181, 187, 191, 194
Warrin, Jeff (see silt)
Wasted Youth, 185
Waters, John, 192
Wavelengths, 214
Werkstattkino, 197
wet-gate printing, 70
Wilke, Hannah, 164
Williams, Walter, 136
Wittner, 61, 104
Wollensak, 184
Wood, Robin, 188
work print, 60, 68, 70, 143
Würker splicer, 87
Yashica, 16, 160, 162, 184
Yashica SU-40E, 160
Yashica Super 600 Electro, 43
Year of the Horse (see Jarmusch, Jim and Young, Neil)
York University, 187, 188
Young, Neil (*Greendale, Year of the Horse*), 61
YouTube, 28, 218, 220, 221, 222, 224
Zavada, Roland, 50, 56, 98, 138-141
ZE Records, 166
Zedd, Nick, 20
Zeiss Ikon Movipress-Super 8, 83
Zeiss Moviscop S8, 83
Zenit Quarz 1x8C-2, 43
Zilkha, Micael, 166
Zoom, 16, 32, 34, 46, 144, 202

THANK YOU

Alison Levy for reading countless versions of this book, and serving as an early-stages copy editor, as well as coming in with a title at the last minute.

Mark Taylor for editing this book, as well as lots of critical thinking around the project.

Victor Krummenacher and Daphnée Branchy for their beautiful book design.

Jeremy Troy for creative direction and consultation.

Much technical advice and expertise was sought along the way. Dino Everett, archivist from the USC SCA Hugh M. Hefner Moving Image Archive, was critical on this front. He also was instrumental in scanning countless images and films for my use. I interviewed a lot of folks for this book, many of whom are granted full interviews within its pages. Phil Vigeant of Pro8mm is one of those guys, but I need to send a shout-out to Rhonda Vigeant as well for all her help. Jennifer Miko and Buck Bito of Movette Film Transfer were incredibly helpful answering all of my queries, and much of the equipment photographed belongs to them. Others whom I consulted include Keith Anderson of Yale Labs, Ross Lipman, Michael Anders, Jeff Kreines, Bob Shanebrook, Gibbs Chapman, and Roland Zavada, who was always ready to answer one more question.

The majority of the photography in this book was shot in a manic three days. Jeremy Troy and Colin James Russell were instrumental in designing the lighting for the shoot. Mr. Troy and Joanne Chu were instrumental in touching up the photographs in post. I must also thank Joanne for her behind-the-scenes design work. Photography assistants from the University of San Francisco included Miles Herman, Ellie Vanderlip, Sophie Schwarz, Karen Ver Trinidad, Alexander Flores, and Riley Evans. They assisted with lights and cleaned and prepped all the gear, which was no easy task given the state of most of the equipment. Thanks to Jan Frei for opening the doors at Teak to allow us to take over the studio for a whirlwind weekend. The balance of the photographs were shot at the University of San Francisco.

Kodak was incredibly supportive of this book. Rich Tavtigian put me in touch with whomever I needed, whenever I needed. Others who were a big help at Kodak include Joshua Robertson, James Manelis, Bill Herman, Laura Zigarowicz, Joshua Coon, and James Bulmahn.

The University of San Francisco provided research and equipment funds, research students, and a framework that allowed me to write this book. Thank you to Deborah Benrubi, Hwa-Ji Shin, Joshua Gamson, Charlotte Roh, Sheri Brenner, Chris Brooks, the university's Faculty Development Fund, and the Writing Groups program at USF. Research assistants include Miles Herman, Mia Dixon-Slaughter, Erin Abbatiello, and Riley Kam.

I dug through a lot of archives. I appreciate the efforts of Jesse Peers, Todd Gustavson, Sophia Lorent, Kathy Connor, Lauren Lean, and Stephanie Hofner at the George Eastman Museum; Miranda Mims & Melinda Wallington at the University of Rochester; Mark Toscano at the Academy of Motion Picture Arts and Sciences Film Archive; Amy Sloper at the Harvard Film Archive. Thank you to Paul Sheptow of *Super-8 Filmmaker* for opening up and gifting me your archive of materials. Additional historical research help came from Frank Uhle, Reinhard Wolf, Gerry Fialka, and Seth Mitter and Antonella Bonfanti at Canyon Cinema.

Lots of other images and equipment arrived courtesy of Christian Bruno, Frank Bruinsma, Jim Granato, Donald Deveau, Mark Brecke, Roger Beebe, Ludwig Draser, Elizabeth Pepin Silva, Ken Paul Rosenthal, Deb Pastor, David Schendel, Alan Petty, Abigail Severance, Ignacio Benedeti Corzo, and Daniel Nicoletta. Not all of the material made it onto the page, but all these contributions are valued.

More big thanks go to Henry Plotnick, Rich Hyatt, Anthony Bedard, Bill Daniel, Brecht Andersch, Frank Martinez, Alec Rodriques, Veronica Wolff, Beth Lisick, Paul Haar, Gail Silva, and Sharon Silva.

Special thanks to Ronnie Garver, who designed an early draft of this book that inspired our final design.

At Rare Bird, I would like to thank Tyson Cornell, Julia Callahan, Hailie Johnson, Daphnée Branchy, and Guy Intoci.

THIS IS A GENUINE RARE BIRD BOOK

A Rare Bird Book | Rare Bird Books
453 South Spring Street, Suite 302
Los Angeles, CA 90013
rarebirdbooks.com

Copyright © 2020 by Danny Plotnick

Photographs used with permission.
All rights reserved:

FIRST HARDCOVER EDITION

All rights reserved, including the right to reproduce this book
or portions thereof in any form whatsoever,
including but not limited to print, audio, and electronic.

For more information, address:
Rare Bird Books Subsidiary Rights Department
453 South Spring Street, Suite 302
Los Angeles, CA 90013

Set in Avenir Next
Printed in China

Book Design by Victor Krummenacher

10 9 8 7 6 5 4 3 2 1

Publisher's Cataloging-in-Publication Data
available on request.